ULSI Front-End Technology

Covering from the First Semiconductor Paper to CMOS FINFET Technology

ULSI Front-End Technology

Covering from the First Semiconductor Paper to CMOS FINFET Technology

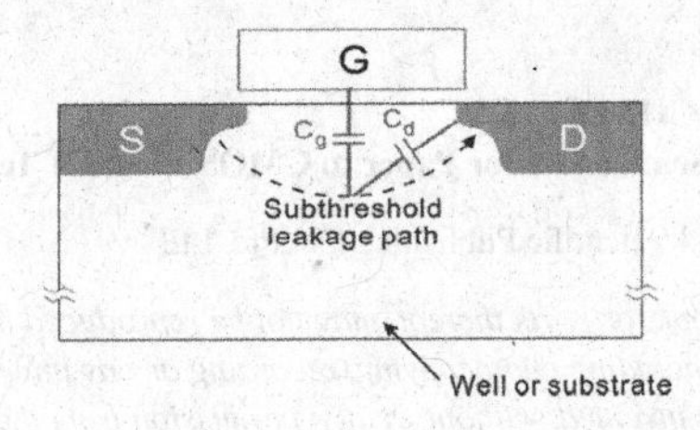

W S Lau

NEW JERSEY • LONDON • SINGAPORE • BEIJING • SHANGHAI • HONG KONG • TAIPEI • CHENNAI • TOKYO

Published by

World Scientific Publishing Co. Pte. Ltd.
5 Toh Tuck Link, Singapore 596224
USA office: 27 Warren Street, Suite 401-402, Hackensack, NJ 07601
UK office: 57 Shelton Street, Covent Garden, London WC2H 9HE

British Library Cataloguing-in-Publication Data
A catalogue record for this book is available from the British Library.

ULSI FRONT-END TECHNOLOGY
Covering from the First Semiconductor Paper to CMOS FINFET Technology

ISBN 978-981-3222-15-1

Desk Editor: Rhaimie Wahap

Typeset by Stallion Press
Email: enquiries@stallionpress.com

Printed in Singapore

Contents

Preface

Since 1977, the author has been working on electronic materials and devices. Previously, he has published a book — *Infrared Characterization for Microelectronics*. The book was published by World Scientific Publishing Company in 1999. The focus of the book was more on electronic materials. After the publication of the book, the author then had the idea to publish a book on electronic devices and integrated circuits. The book was finally completed in 2016.

The title of this book is *ULSI Front-end Technology*. ULSI is short for ultra large scale integration. Front-end technology is the technology of the active devices like n-channel and p-channel MOS transistor in CMOS technology. Back-end technology is the technology of the interconnection of active devices. Back-end technology is equally important as it can seriously affect the yield of integrated circuits. Originally, this book was planned to include back-end technology, to be written by the author's co-worker. However, the co-worker gave up on the writing. In the present version of this book, there is still some discussion on back-end technology. To avoid copyright issues many of the figures have been re-drawn.

This book has been planned to include the very early history of solid state devices and integrated circuits. The first scientific paper on semiconductor written by Karl Ferdinand Braun (1850–1918) in 1874 is discussed in Chapter 1. The invention of p-n junctions, bipolar transistors and MOS transistors is also briefly discussed in the same chapter. Subsequently in Chapter 2, the history of PMOS, NMOS and then CMOS integrated circuit technology is covered. The basic theory of high speed and low power integrated circuit technology is discussed in detail in Chapters 3 and 4 respectively. Besides digital integrated circuit technology, analog integrated circuit technology is discussed in Chapter 5. The combination of digital and analog functions in the same integrated circuit is known as mixed-signal technology; this is also included in Chapter 5. The present state-of-the-art silicon-based FINFET technology is discussed in Chapters 3–5.

The author has been working on high-k dielectric materials for about 30 years. He has published scientific papers on tantalum oxide, aluminum oxide and hafnium oxide thin films. The deposition of these thin films was by sputtering, chemical vapor deposition (CVD) or atomic layer deposition (ALD). In 2012, he has published his theory regarding leakage current mechanism in high-k dielectric materials. Some of his works on high-k MIM (metal-insulator-metal) capacitors are included in Chapter 5. Historically, leakage current mechanism in high-k dielectric materials is full of confusion and contradictions. New insights on high-k dielectric materials from the author can be found in this book. Essentially, this book covers the historic period from 1874 to 2016.

Author Biography

Dr. W. S. Lau was born in Hong Kong in 1955. He got his PhD in Electrical Engineering from the Pennsylvania State University (USA) in 1987. He served in Chartered Semiconductor Manufacturing Ltd. (Singapore) from July 1997 to January 2001 and then joined the Division of Microelectronics, School of Electrical and Electronic Engineering, Nanyang Technological University, Singapore. He has been working on Si-based CMOS technology from 1997 onwards. On the whole he is more familiar with the front end of line (FEOL) technology than the back end of line (BEOL) technology. He has been working on both Si-based and III-V semiconductor based materials and devices for many years. For constructive feedback, the author may be reached by e-mail as follows:

W. S. Lau
lauwaishing@yahoo.com.sg

Chapter One

Introduction to the History of Semiconductors

1.1 Early History of Semiconductors

According to an article regarding the very early history of semiconductors written by Georg Busch,[1] the first person who mentioned something close to the word "semiconductor" was an Italian scientist Alexandro Volta (1745–1827), who was famous for the invention of the electric battery. Volta was born in Como, Lombardy, Italy. However, during his life span, a unified Italy did not exist; Volta was first under the rule of the Emperor of Austria, then under the rule of Napoleon Bonaparte and then under the rule of the Emperor of Austria again. It is interesting to note that Volta published in the Philosophical Transactions of the Royal Society of London, which is the oldest scientific journal in the English-speaking world, in the United Kingdom since 1665. In 1782, Volta published a paper in Philosophical Transactions of the Royal Society of London.[2] It was unclear if he wrote his paper in Italian or in French; however, at the end of his paper, an English translation appeared. Inside the English translation, a small passage can be found as follows. "The surface of those bodies does not contract any electricity, or if any electricity adheres to them, it vanishes soon, on account of their *semi-conducting* nature; for which reason they cannot answer the office of an electrophorus, and therefore are more fit to be used as condensers of electricity."

Humphry Davy (1778–1829) was a famous UK scientist who served as the President of the Royal Society from 1820–1827. He discovered chlorine and iodine. In 1821, Davy mentioned his observation of the effect of increasing temperature on the electrical conductivity of metals as follows.[3] "The most remarkable general result that I obtained by these researches, and which I shall mention first, as it influences all others, was, that the conducting power of metallic bodies varied with the temperature, and was lower in some inverse ratio as the temperature was higher." Michael Faraday (1791–1867) was an English chemist and physicist. He contributed significantly to the understanding of electromagnetism and electrochemistry. Faraday's experimental work in chemistry led him to the first documented observation of a material which is now known as a semiconductor. In 1833, he found that the electrical conductivity of silver sulfide increased with increasing temperature as follows.[4] "The effect of heat in increasing the conducting power of many substances, especially for electricity of high tension, is well known. I have lately met with an extraordinary case of this kind, for electricity of low tension, or that of the voltaic pile, and which is in direct contrast with the influence of heat upon metallic bodies and decribed by Sir Humphry Davy." In his 1833 paper, Faraday mentioned Davy's 1821 paper. In fact, Michael Faraday once served as Davy's assistant. "The substance presenting this effect is suphuret of silver. It was made by fusing a mixture of precipitated silver and sublimed sulphur, removing the film of silver by a file from the exterior of the fused mass, pulverizing the sulphuret, mingling it with more sulphur, and fusing it again in a green glass tube, so that no air should obtain access during the process. The surface of the sulphuret being again removed by a file or knife, it was considered quite free from uncombined siliver." For a metal, the electrical conductivity decreases with increasing temperature. Sulphuret of silver is now known as silver sulfide (Ag_2S), which is a direct bandgap semiconductor with a bandgap of about 1 eV.[5] Thus this effect usually found in semiconductors, is opposite to the situation usually found for metals.

With the help of modern physics, we can easily see that raising the temperature of most semiconductors increases the density of free carriers (electrons or holes) inside them and hence their conductivity. This effect can be exploited to make thermistors whose resistance is sensitive to a change in temperature. For metals, the density of free carriers (electrons) is not influenced by a higher temperature; however, a higher temperature implies stronger scattering and thus lower electron mobility. For semiconductors, a higher temperature also implies stronger scattering and thus lower electron mobility but the increase in carrier density due to higher temperature can be the stronger and thus the dominant effect. Thus Faraday's 1833 paper can be considered as the first scientific paper loosely related to semiconductor physics.

For semiconductors, the behavior of an extrinsic semiconductor (a semiconductor doped by a shallow donor or acceptor) is similar to that of metals reported by Davy; however, the behavior of an intrinsic semiconductor (for example, an undoped semiconductor) is similar to that of silver sulfide reported by Faraday. Thus a resistor made of an extrinsic semiconductor can show up a positive temperature coefficient of resistance while a resistor made of an intrinsic semiconductor can show up a negative temperature coefficient of resistance.

For modern MOS transistors, the carrier mobility involves 3 scattering mechanisms: Coulombic scattering, phonon scattering and surface roughness scattering. All 3 scattering mechanisms become stronger at higher temperature, resulting in lower mobility at higher temperature.[6] For modern MOS transistors, the on current decreases with increasing temperature just like Davy's report while the off current increases with increasing temperature in a way similar to an intrinsic semiconductor and thus similar to Faraday's observation. An interesting phenomenon is that for MOS transistors there exists a cross-over point where the current is insensitive to temperature variation.[7–11] As shown in Fig. 1.1, the drain current versus gate voltage characteristics of an n-channel MOS transistor show up a TIP at a particular value of gate voltage. TIP stands for "temperature independent point". For gate voltage above TIP, the drain current decreases with temperature. For gate voltage below TIP, the drain current increases with temperature. Similarly, as shown in Fig. 1.2, the drain current versus gate voltage characteristics of a p-channel MOS transistor show up a TIP at a particular value of gate voltage. Thus a resistor made up of the drain and source electrodes of an MOS transistor with a fixed gate-to-source voltage

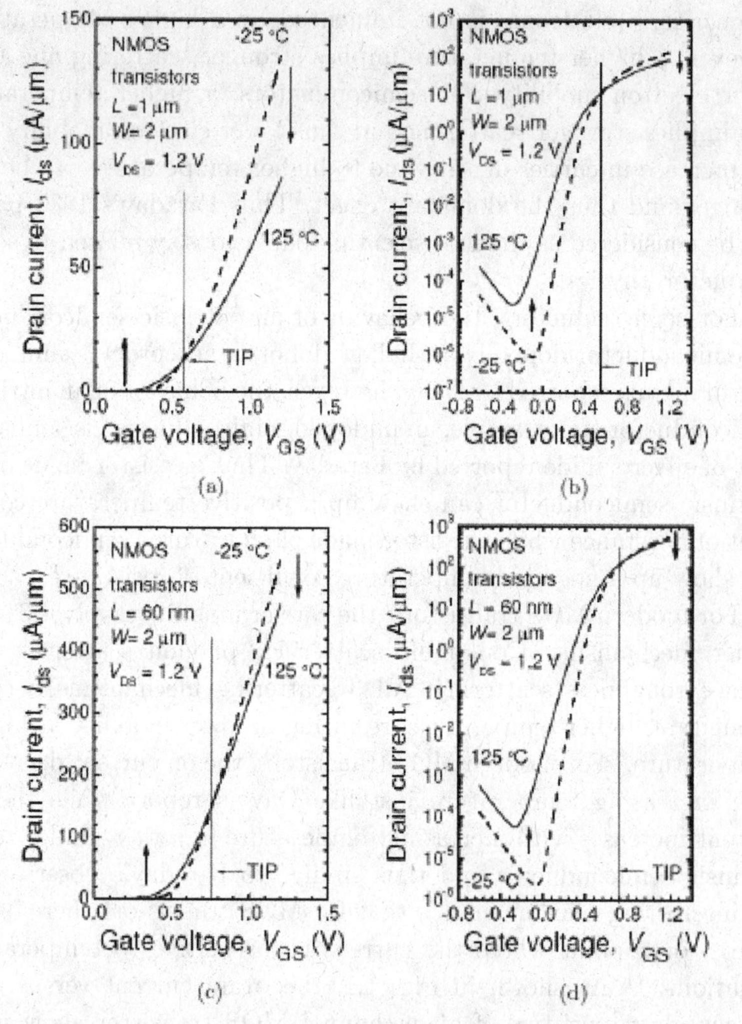

Fig. 1.1 Effects of temperature on the on-state current (/on) and off-state current (/off) of NMOS transistor according to Yang *et al.*[a,11]

can show up a positive temperature coefficient of resistance above TIP and a negative temperature coefficient of resistance below TIP, respectively. Another conclusion can be drawn from Fig. 1.1 and

[a]P. Yang was a PhD student of the author and so this figure is part of the work of the author.

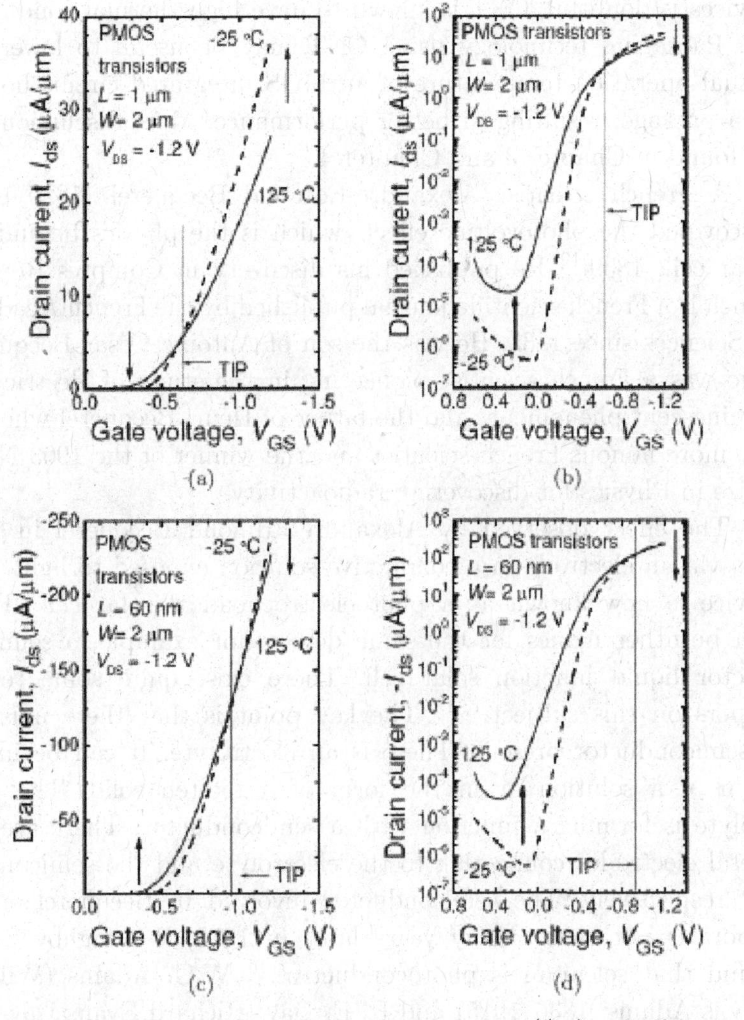

Fig. 1.2 Effects of temperature on the on-state current (/on) and off-state current (/off) of PMOS transistor according to Yang *et al.*[a,11]

Fig. 1.2: operation of MOS transistors at low temperature implies higher on current and lower off current; the implication is that MOS technology usually improves by a decrease in operation temperature. For example, carbon nanotube (CNT) technology has been shown that device operation is possible. However, manufacturability of CNT devices is doubtful. CNT is known to have high thermal conductivity. Packaging technology using CNT may be useful to lower the actual operation temperature of an MOS integrated circuit housed in a package, resulting in better performance. More discussion can be found in Chapter 3 and Chapter 4.

A French scientist Alexandre-Edmond Becquerel (1820–1891) discovered the photovoltaic effect, which is the physics behind the solar cell, 1839.[12] He published his discovery in Comptes Rendus, which is a French scientific journal published by the French Academy of Sciences since 1835. He was the son of Antoine César Becquerel, who was a French scientist pioneering in the study of electric and luminescent phenomena, and the father of Henri Becquerel who was the more famous French scientist and the winner of the 1903 Nobel Prize in Physics for discovering radioactivity.

The effect observed by Alexandre-Edmond Becquerel in 1839 was via an electrode in a conductive solution exposed to light. The device is now known as a photoelectrochemical solar cell. There can be other names for the same device, for example, a semiconductor liquid junction solar cell. There exist quite some review papers on this subject.[13,14] The key point is that there must be a semiconductor present. There is an electrolyte. It can be in the form of a solution or in the form of a molten solid. The electrolyte is forming a junction with a semiconductor. There are two metal electrodes connecting to the electrolyte and the semiconductor respectively. The semiconductor involved in Becquerel's 1839 report is not clear. Many years later in 1873, Willoughby Smith found that selenium is photoconductive.[15] W.G. Adams (William Grylls Adams, 1836–1915) and R. E. Day (Richard Evans Day, student of Adams) observed the photovoltaic effect in selenium without involving any liquid and reported their observation in 1877.[16] The American Charles Fritts developed a solar cell using selenium with a thin layer of gold in 1883.[17] Thus selenium may be considered the first semiconductor in the solid state discovered by mankind. However at that time both Becquerel and Smith did not know semiconductor physics and thus could not really explain what actually happened.

Photoconductivity is just the formation of free electrons and free holes because light can raise an electron from the valence band to the conduction band leaving behind a hole. It takes many years for the physics of the photovoltaic effect to be understood. Kurt Lehovec (1918–2012) may be the first scientist who can claim that he managed to explain the photovoltaic effect in 1948. Lehovec was born in 1918 in Ledvice, northern Bohemia, Austria-Hungary just a few months before the end of World War I. After the end of World War I in November 1918, Austria-Hungary broke up into several countries. One of these new countries was known as Czechoslovakia with Prague as its capital. According to his own website (www.kurtlehovec.com), after he graduated from high school in 1936, he moved to Prague, where he attended university and received his PhD in Physics in 1941. However, Czechoslovakia was annexed by Nazi Germany in the period of 1938–1939 and Kurt Lehovec became involved in the history of Germany. (Note: Lehovec's website is no longer available after his death in 2012.) At the end of World War II, he spent the next two years in postwar West Germany, continuing with theoretical research and came up with the explanation of the solar cell effect. The US Signal Corps had become aware of his discovery at Prague, and invited him in the summer of 1947 to the USA under the Project Paperclip for outstanding European scientist from the former Nazi-occupied territories. In this way, he ended up in USA. In 1948, he published his theory regarding the photovoltaic effect in the prestigious US journal Physical Review.[18] Besides the photovoltaic effect, he is also known as the inventor of pn junction isolation which is important for integrated circuit technology. He is also known for proposing the theory behind the functioning of light emitting diodes. Many years later, Czechoslovakia broke up into two countries (Czech Republic with Prague as capital and Republic of Slovakia with Bratislava as capital) in 1993. Ledvice, the birth place of Lehovec, is now part of the Czech Republic.

Willoughby Smith (1828–1891) was born in Great Yarmouth, England and was an electrical engineer working on underwater telegraph cable projects. In 1849, he supervised the manufacture and laying of 30 miles of underwater telegraph cable from Dover, which is a town in England facing France across the narrowest part of the English Channel, to Calais, which is a town in France opposite to Dover. He worked closely with Charles Wheatstone (1802–1875) who had designed the machinery for making and laying the cable. (Note: The Wheatstone bridge was named after Charles Wheatstone.) In 1873, Smith developed a method for continually testing an underwater cable as it was being laid. For his test circuit, Smith needed a material with very high resistance but not a complete insulator. He selected selenium rods for this purpose. It turned out that selenium showed some sort of unstable resistance. Joseph May was one of Smith's assistants and he noticed that the resistance of selenium seemed to depend on the amount of light shining on it. The selenium was placed in a box with a sliding cover. With the cover closed, the resistance was highest but it dropped when the cover was open such that light could shine on the selenium. Smith published an article with the title "Effect of light on selenium during the passage of an electric current" in the February 20, 1873 issue of the scientific journal Nature.[15] Nowadays, this phenomenon is commonly known as photoconductivity and selenium is considered a semiconductor. However, at that time, the term "semiconductor" did not exist yet. Selenium was in fact the first elemental semiconductor known to the scientific community. It has a bandgap of about 2 eV. Once upon a time, selenium was commonly used in photocells and rectifiers. Besides photocells and rectifiers, selenium also played a big role in xerography. In the mid 1930's, Chester Floyd Carlson (1906–1968) invented a new method of copying images onto paper. Prior to that time, only a wet process similar to photography was used to make copies. Carlson developed a dry process that transferred powdered ink, called toner, from an optically induced image on a negatively charged transfer device to a piece of positively charged paper. The toner was then heated which melted and fused it onto the paper. In the process of perfecting his new copying technology, Carlson experimented with belts and plates as the transfer device. Neither of these worked very well. Finally, Carlson set upon the idea of using a coated drum as the transfer device. At first, in an attempt to imitate the photographic process, he used silver compounds to coat the drum. On 22nd Oct.

1938, Carlson and his assistant Otto Kornei (1903–1993) had their historic breakthrough using a zinc plate with a sulfur coating.[19] However eventually, Carlson chose selenium as the coating. The process, now known as xerography (from the Greek for "dry writing"), earned Carlson a patent — and now everyone knows about Xerox copiers. Later, when Carlson was asked to identify the most difficult part about inventing xerography, he pointed out that the hardest problem was finding the proper coating. He said, "The primary reason that we settled on selenium is its unique crystal lattice and the way that it retains an electrostatic charge indefinitely. We chose selenium because it truly is the element that never forgets." Now, on the whole, the significance of elemental selenium as a semiconductor has been largely superseded by other semiconductors. Even for xerography, selenium has been challenged by other materials like organic photoconductor (OPC) and amorphous silicon. OPC is cheaper while amorphous silicon is mechanically stronger and so more durable. Besides serving as a photoconductor, selenium can be used for rectifiers. Selenium rectifiers were subsequently more or less replaced by silicon rectifiers. Nowadays (at least up to 2009), some companies can still supply selenium rectifiers. Universal Rectifiers Inc. (Rosenberg, Texas, USA) has a full line of replacement selenium rectifier available. (http://www.universalrectifiers.com/) Cougar Electronics Corp. (New Haven, Connecticut, USA) is another selenium rectifier supplier. (http://www.cougarelectronics.com/)

In 1874, Karl Ferdinand Braun (1850–1918) published a paper in the *Annalen der Physik und Chemie* with the title "Uber die stromleitung durch schwefelmetalle".[20] Annalen der Physik und Chemie was once a very prestigious German journal such that many famous scientists, including Albert Einstein (1879–1955), published their papers in it. An English translation of this paper can be found in the book *Semiconductor Devices: Pioneering Papers*, edited by S. M. Sze. The title, after translated into English, is "On current conduction through metallic sulfides". Braun was a prominent German scientist who shared the 1909 Nobel Prize in Physics with the Italian Guglielmo Marconi (1874–1937) for their contributions to the development of wireless telegraphy. Braun noticed that the current flow can be influenced by the voltage polarity. For materials obeying Ohm's Law, the current-voltage (I-V) characteristics is both linear and symmetric. Thus Braun reported the existence of non-linear asymmetric current-voltage (I-V) characteristics for some metallic sulfides, which are semiconductors according to the modern sense. For example, lead sulfide (PbS) is a semiconductor with a bandgap of about 0.4 eV.[21] Galena is a naturally occurring mineral form of lead sulfide. Once upon a time, a cat's whisker detector based on galena was used in primitive radios. Thus Braun's 1874 paper can be considered the first paper on semiconductors while Faraday's 1833 paper can be considered as the first paper somewhat loosely related to semiconductors.

Braun only reported an interesting phenomenon. Eventually an application was found. Amplitude modulation (AM) is one of the many possible ways to modulate radio waves for wireless communication. In the early days of wireless, a cat whisker detector may be used in very primitive AM radio receivers. It is simply a relatively primitive and unstable metal–semiconductor point-contact junction forming a Schottky barrier diode. One side of the diode is a metal wire. The other side of the diode is a semiconductor. The key point that such a device can be used to demodulate AM radio signal is the asymmetric current-voltage (I-V) characteristics first reported by Braun in 1874. The theory to explain the asymmetric I-V characteristics was subsequently proposed by another German scientist Walter Schottky (1886–1976).[22] There are many possible choices for the semiconductor. As discussed above, galena (PbS, with a bandgap of about 0.4 eV) can be a possible choice. For example, Sir Jagadish Chandra Bose (1858–1937), who was a Bengali scientist born under British rule, got a patent regarding the use of galena and some other materials for radio detector applications.[23] It was Braun who made an important observation but it was Sir J.C. Bose who thought out a practical application. The discovery of the element germanium is usually attributed to a German chemist Clemens Alexander Winkler (1838–1904). A prominent Russian chemist Dmitri Ivanovich Mendeleev (1834–1907) predicted the existence of germanium through the periodic table of elements; however, it was Winkler who discovered germanium in 1886 and named the element after his own country, Germany. Silicon was discovered before germanium. The discovery of the element silicon is usually attributed to a Swedish chemist Jons Jacob Berzelius (1779–1848). In 1824 Berzelius prepared amorphous silicon by heating potassium with silicon tetrafluoride (SiF_4). A German chemist Friedrich Wohler (1800–1882) managed to prepare silicon in a crystalline form. One of the earliest applications of silicon is to add silicon to steel, resulting in a material known as silicon steel which can be used in the iron core of electrical transformers with less eddy current loss. The invention of silicon steel is usually attributed to Sir Robert Abott Hadfield (1858–1940), who was an English metallurgist. In 1906, Greenleaf Whittier Pickard (1877–1956), who was a US radio pioneer, filed a patent for a silicon based radio detector.[24] Nowadays, PbS is no longer used for AM radio detector but PbS or PbSe infrared detectors are still commercially available. For example, Teledyne Judson Technologies (Montgomeryville, Pennsylvania, USA) is still selling PbS infrared detectors for operation in the 1–3.5 μm wavelength region. Hamamatsu is another PbS infrared detector supplier. Primitive Si point contact detector is no longer used for AM radio detector but modern Si point contact detectors are still used at UHF or microwave frequencies. For example, Advanced Semiconductor Inc. (North Hollywood, California, USA) is still selling Si point contact detectors workable up to 16 GHz.

Subsequently, this kind of primitive radio detectors were almost totally replaced by vacuum tube diode detectors. Sir John Ambrose Fleming (1849–1945) was usually considered the scientist who invented vacuum tube diodes; he was a prominent English scientist and he got US Patent 803,684 in 1905 for this invention.[25] The basic principle of the operation of the vacuum tube diode was discovered by Frederick Guthrie (1833–1886), who was a British scientific writer and professor, in 1873. Subsequently, the US inventor Thomas Alva Edison (1847–1931) independently rediscovered it in 1880 and the effect came to be known as "Edison effect". However, it was Sir Fleming who thought out a practical application. Vacuum tube diodes can be used as rectifiers and also as AM radio detectors.

Primitive semiconductor radio detectors had difficulty to compete with vacuum tube diode detectors. Both semiconductor and vacuum tube diode detectors cannot amplify electrical signals. This situation was changed by the invention of vacuum tube triodes, tetrodes and pentodes. Lee de Forest (1873–1961) was an American; he was usually considered the scientist who invented vacuum tube triodes and he got US Patent 879,532 in 1908 for this invention.[26] There is some controversy regarding who really invented the vacuum tube triode; in fact, there is a claim that an Austrian physicist Robert von Lieben (1878–1913) invented something similar. Walter Schottky (1886–1976), who was a prominent German physicist, invented the vacuum tube tetrode; he received German patent 300, 617 in 1916 for this invention. Bernard D. H. Tellegen (1900–1990) was a prominent Dutch scientist and engineer; he was usually considered the scientist who invented vacuum tube pentodes and he got US Patent 1,945,040 in 1934 for this invention.[27]

AM radios based on primitive semiconductor diode detectors simply cannot compete with AM radios based on vacuum tubes. However, the need to develop microwave radar for aircraft detection before and during World War II gave semiconductor diode detectors a new life. This is because vacuum tube diode detectors have difficulty to operate at the microwave frequencies required for radar. The size of the antenna required for radar can be decreased by using higher frequencies. At frequencies of the order of GHz, it was found that the old cat's whisker detector, which is actually a primitive semiconductor diode detector, can perform better than a vacuum tube diode detector, resulting in renewed interest in the old cat's whisker detector. Robert Buderi published a book regarding the invention of radar. He also mentioned that radar technology helped to stimulate the development of semiconductor technology in Chapter 15.[28]

John Orton (Emeritus Professor, School of EEE, University of Nottingham, UK) gave a good explanation regarding this in his book.[29] The depletion region in an old cat's whisker detector is very small and so the transit time is very small compared to a vacuum tube diode. The potential harmful effects of a finite transit time in vacuum tubes were discussed by various authors, for example, Llewellyn

(Bell Laboratories).[30] In addition, the old cat's whisker detector is basically a point contact detector with a very small area and so the capacitance is very small.

1.2 Invention of the p-n Junction

However, the old cat's whisker detector can be very unstable and the reproducibility can be very poor. For example, a "good" spot has to be found. One of the best choices for the semiconductor in a cat's whisker detector was found to be silicon. It was also observed that the detected signal can have two polarities; nowadays, this can be understood because silicon can be p-type or n-type. A lot of effort has been spent on semiconductor research for military applications during World War II. For example, Frederick Seitz (1911–2008) published an article in Physics Today regarding the research done on silicon and germanium during World War II.[31] Russell Shoemaker Ohl (1898–1987) and his co-workers in the Bell Laboratories managed to make p-type silicon, n-type silicon in a reproducible manner. In fact he made the first p-n junction in silicon as follows. In 1947 after World War II, Scaff and Ohl published a report on their effort to develop silicon microwave detectors.[32] Morrison[32] demonstrated the superiority of silicon detectors compared to vacuum tube detectors at microwave frequencies by actual measurement.

Ohl was an American working in the Bell Laboratories. He was quite frequently considered an unsung hero in the semiconductor revolution. Proper credit to Ohl was given by Riordan and Hoddeson in their 1997 article in IEEE Spectrum[33] and their book "Crystal Fire".[34] In 1940, Ohl was working with a silicon crystal sample that had a crack down its middle. He was using an ohmmeter to test the electrical resistance of the sample when he noted that when the sample was exposed to light, the current that flowed between the two sides of the crack made a significant jump. It was known that other semiconductors, such as selenium (Se), generated a small current when exposed to light. However, the cracked silicon sample was quite a curiosity. Ohl showed the sample to his colleagues in the Bell Laboratories and together they deduced that the crack was a fortunate accident: It marked the dividing line that had occurred when the molten silicon froze in the crucible. At that moment, various impurities or contaminants in the silicon had been isolated into different regions, with the crack separating them. As a result, the silicon atoms in the region on one side of the crack had extra electrons around them. The other region was the opposite; its crystallized silicon had a slight shortage of electrons. They named the two regions p and n: p for positive-type and n for negative-type. The barrier between the impurities was called the p-n junction. The junction represented a barrier, preventing the excess electrons in the n-region from traveling over to the p-region which is short of electrons, resulting in zero current. However, when the sample was irradiated by light, there is a current flow, resulting in a simple device which can convert light into electrical energy. Thus Ohl invented the silicon p-n junction solar cell; he got US Patent 2,402,662 in 1946 for this invention.[35] Unlike the earlier selenium solar cells, the silicon solar cells based on the p-n junction converted sunlight much more efficiently. Ohl also got US Patent 2,402,662 in 1946 for his invention of the p-n junction.[36](Note: This phenomenon was subsequently studied in much more detail by scientists. When the silicon melt has impurities, the impurities will be distributed between the silicon solid crystal and the silicon melt during the solidification process. A lot of work has been done on how impurities are incorporated into the silicon crystal during the crystallization process. If the silicon melt has both p-type and n-type impurities present by accident, it is possible that, after solidification, part of the silicon becomes p-type and another part of the silicon becomes n-type, resulting in a p-n junction unintentionally. So far the best explanation came from the book "The story of semiconductors" by John Orton as follows. The silicon melt contained both boron and phosphorus impurities with significantly different segregation coefficients "k". Phosphorus atoms ($k = 0.04$) were swept to the bottom of the freezing silicon ingot, doping it n-type while boron atoms ($k = 0.8$) tended to remain fairly uniformly distributed. At the top of the solidified silicon ingot, the concentration of boron was greater than that of phosphorus, resulting in p-type behavior, while the converse was true at the bottom of the solidified silicon ingot. The readers should note that the segregation coefficients quoted by Orton may not be the same as those numbers published in other books or journal papers. The effective segregation coefficient or distribution coefficient is not a constant but it depends on various parameters like the crystallization speed, rotation speed, etc. Nevertheless, Orton's explanation is the best one found by the author.)

According to Buderi,[28] it was two Bell Laboratories workers: Jack Theuerer and Henry Scaff who theorized that p-type conductivity resulted from trace elements like boron from the third column of the periodic table while n-type conductivity came from trace elements like phosphorus from the fifth column of the periodic table. In fact, Scaff and Theuerer published a paper in 1949 regarding this.[37] However, other scientists might make a similar claim regarding the invention of semiconductor doping. The invention of doping of semiconductors was quite frequently attributed to John Robert Woodyard (1904–1981). He got a patent on the doping of germanium as US Patent 2,530,110, which was filed in 1944 and awarded in 1950.[38]

1.3 Invention of the Transistor

Both the old cat's whisker detector and the p-n junction diode detector cannot amplify electrical signals. After World War II, John Bardeen (1908–1991) and Walter Houser Brattain (1902–1987) invented the point-contact transistor in the Bell Laboratories.[39–41] William Bradford Shockley (Bell Laboratories) invented the junction transistor.[42,43] William Bradford Shockley (1910–1989), John Bardeen (1908–1991) and Walter Houser Brattain (1902–1987) received the 1956 Nobel Prize in physics for their discovery on the transistor effect. It was John Robinson Pierce (1910–2002) who, at the request of Brattain, coined the name "transistor". At that time, Pierce was the supervisor of the Bell Laboratories transistor team. Shockley subsequently left Bell Laboratories to start up his own company with the name of the Shockley Transistor Company, which did not survive long. Currently there is a website (www.shockleytransistor.com) dedicated to the memory of Shockley and his company. 8 important members of Shockley's company, including Jean Amedee Hoerni (1924–1997), Gordon Earle Moore (1929–) and Robert Norton Noyce (1927–1990), quitted to join Fairchild Semiconductor. Jean Hoerni was famous for the development of the planar process. Gordon Moore and Robert Noyce have been famous as two of the three founders of Intel.

Besides US scientists, it is also known that two German physicists Herbert Franz Mataré (1912–2011) and Heinrich Welker (1912–1981), who was working in France after World War II, independently also developed something similar to a transistor in around 1948, roughly at the same time and independently from the Bell Labs engineers.[44,45] Mataré returned to Germany and in 1952 co-founded Intermetall to manufacture diodes and transistors. Welker joined Siemens, eventually becoming its research director. He is also remembered for performing fundamental research on III-V semiconductors. In 1952, he described semiconductors from elements found in column III and V of the periodic table as potentially useful for electronic devices. One of these, gallium arsenide (GaAs) was to feature prominently in the search for an efficient communication laser. Thus Welker failed to be remembered as the inventor of transistor but he is remembered as the scientist who recognized the potential of III-V semiconductors. There exists another claim that a female scientist Nina Aleksandrovna Goryunova (1916–1971) of USSR also recognized the potential of III-V semiconductors. Goryunova described III-V materials as semiconductors for the first time in 1950. In her Ph.D. dissertation, completed in 1951 at Leningrad State University (now known as Saint Petersburg State University), she indicated that III-V compounds with the zinc-blende (cubic zinc sulfide, ZnS) crystalline structure are semiconductors. Her work was not published outside of the USSR until much later due to the Cold War. However, Welker seems to be the more famous person. As we shall see later in this book, some scientists have been investigating the use of III-V semiconductors in mainstream CMOS integrated circuits.

The first transistors made in the Bell Laboratories were based on germanium. Germanium has a small bandgap of about 0.7 eV. When the temperature increases, leakage current increases exponentially.

Silicon has a larger bandgap of about 1.1 eV. It can be easily seen that silicon transistors will have much better thermal stability compared to germanium transistors. The first successful silicon transistors were made in Texas Instruments. Gordon Kidd Teal (1907–2003) was the leader of the TI team responsible for this feat.[46] He worked for the Bell Laboratories before joining TI. The story was that it was late afternoon at a conference organized by the Institute of Radio Engineers (IRE, now known as IEEE) in 1954. People complained the poor performance of Ge transistors at high temperatures and expected that Si transistors would perform better. However, they believed that viable Si transistors would not be available soon. Suddenly, Teal came out to give his talk. He pulled three small objects out of his pocket and announced: "Contrary to what my colleagues have told you about the bleak prospects for silicon transistors, I happen to have a few of them here in my pocket." TI became the first company to produce silicon transistors. After the TI success to make silicon transistors, scientists working in the Bell Laboratories also managed to make silicon transistors. For example, Tanenbaum and Thomas published a paper "Diffused emitter and base silicon transistors" in 1956.[47] Aschner *et al.* published a paper "A double diffused silicon high-frequency switching transistor produced by oxide masking techniques" in 1959.[48] Theuerer *et al.* published a famous paper on "Epitaxial diffused transistors" in 1960.[49] For discrete silicon transistors, the double diffused epitaxial transistor eventually becomes the standard structure.

1.4 Invention of the Integrated Circuit

Jack St. Clair Kilby (1923–2005) is a Nobel Prize laureate in physics in 2000 for his invention of the integrated circuit in 1958 while working at Texas Instruments (TI).[50] He is quite frequently known as Jack Kilby or J.S. Kilby. Many years later, Kilby published about the history of his invention.[51,52] However, Robert Norton Noyce (1927–1990), who was one of the three founders of Intel, also claimed to have invented the integrated circuit.[53] According to Warner,[54] Bell Laboratories missed the chance to be the champion of integrated circuits.

In 2009, Arjun Saxena, an Emeritus Professor of the Rensselaer Polytechnic Institute (USA), published an interesting book on the invention of integrated circuits.[55]

There are more than one transistor in an integrated circuit and thus device isolation is necessary. Kurt Lehovec (1918–2012) pioneered "junction isolation". He obtained US Patent 3,029,366 for this.[56] Later he published a short paper regarding the history of his invention in 1978.[57] While Lehovec pioneered "junction isolation", Jean Amedee Hoerni (1924–1997) pioneered "planar technology". He obtained US Patent 3,025,589[58] and US Patent 3,064,167.[59] He published a few paragraphs in 1961 on his great invention.[60] The process involves the basic procedures of silicon dioxide (SiO_2) oxidation, SiO_2 etching and doping by thermal diffusion. The final steps involve oxidizing the entire wafer such that a SiO_2 insulating film covers the wafer, etching contact holes to the transistors, and depositing a covering metal layer over the oxide, thus connecting the transistors without manually wiring them together. This major milestone was achieved when Hoerni was with Fairchild. Many years later, Riordan wrote a very interesting article in Spectrum regarding this great invention.[61] As Riordan pointed out, the success of the silicon planar technology depends very much on the existence of a good oxide, silicon dioxide, for silicon. (Note: Silicon has two important oxides: silicon monoxide and silicon dioxide. Silicon monoxide can sublime at relatively low temperature and so not suitable for the planar technology. However, silicon monoxide is popularly used as an anti-reflection coating. Silicon dioxide is the oxide which is important both for discrete transistor and integrated circuit technology.) L. Derick and C. J. Frosch (Bell Laboratories) contributed in this aspect by their patent.[62] They also contributed their paper "Surface protection and selective masking during diffusion in silicon" in 1957.[63]

Figure 1.3 shows the cross-sections of 2 double diffused silicon npn transistors according to Aschener *et al.* (Bell Laboratories).[48] The area of the collector-base junction was defined by a mesa etch process. (Note: The mesa transistor structure was probably developed by engineers in Texas Instruments. For example, Texas

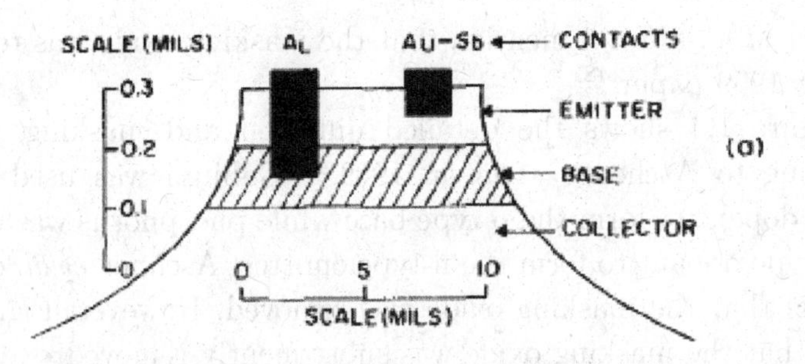

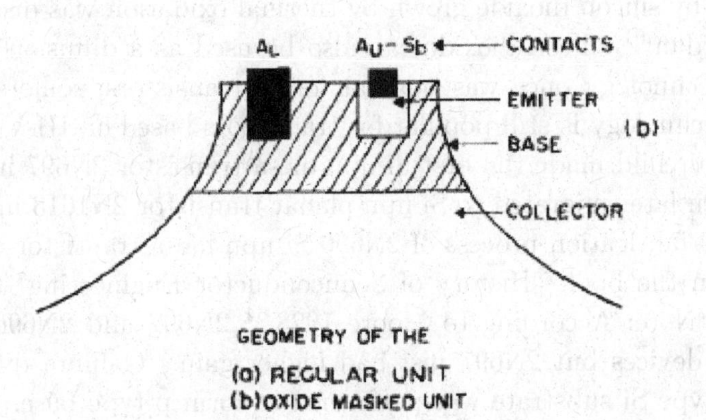

Fig. 1.3 Cross-sections of two double diffused transistor structures: (a) regular unit without oxide masking (b) with oxide masking according to Aschner *et al.* 1959 (Bell Laboratories).[48] (Reprinted with permission from J. F. Aschner, C. A. Bittmann, W. F. J. Hare and J. J. Kleimack, “A double diffused silicon high-frequency switching transistor produced by oxide masking techniques”, *Journal of the Electrochemical Society,* vol. 106, no. 5 (May 1959) pp. 415–417. Copyright 1959 Electrochemical Society. Reproduced by permission of The Electrochemical Society.)

Instruments introduced the 2N389 transistor, which was the first silicon power transistor available to industry and used a mesa structure, in 1957.)

In an article by Moore,[64] a figure shows a mesa transistor made by Fairchild in the early days of silicon technology; it is quite similar to the structure of Aschner *et al.* (Bell Laboratories) 1959. In the transistor structure shown in Fig. 1.3, there is no passivation.

Aschner *et al.* did not mention that the masking oxide was removed in their 1959 paper.[48]

Figure 1.4 shows the detailed diffusion and masking process according to Aschener *et al.* in 1959.[48] Gallium was used as the p-type dopant to form the p-type base while phosphorus was used as the n-type dopant to form the n-type emitter. Aschner *et al.* did not mention that the masking oxide was removed. However, Fig. 1.4(e) shows that the masking oxide was subsequently removed.

Bipolar transistors can be fabricated by mesa technology or by planar technology. Planar technology with passivation of the silicon surface by silicon dioxide grown by thermal oxidation was discussed by Riordan;[61] silicon dioxide can also be used as a diffusion mask. Mesa technology once was popular for Si transistors; some sort of mesa technology is still popular for transistors based on III-V materials. Fairchild made the first Si npn mesa transistor 2N697 in 1958 and then later migrated to Si npn planar transistor 2N1613 in 1960. The full fabrication process of 2N696 Si npn mesa transistor can be found in the book "History of Semiconductor Engineering" by Bo Lojek. (Note: According to Moore 1998,[64] 2N697 and 2N696 were similar devices but 2N697 just had higher gain.) Gallium diffusion into n-type Si substrate was performed to form p-type base. (Note: This is similar to Aschner *et al.* 1959.[48]) This was a blanket diffusion into the whole Si substrate and so there was no base mask. (Note: Silicon dioxide cannot be used to mask gallium diffusion into silicon. Nowadays, people seldom used gallium doping in silicon based microelectronics.) Subsequently, phosphorus diffusion into the p-type region was performed to form n-type emitter. This was done through an opening in a silicon dioxide mask and so there was an emitter mask for this step. Further down the process flow, there was a step to do a mesa etch such that most of the p-type region was etched away except for the p-type base. This mesa etch step controls the area of the collector-base junction. For planar technology, boron diffusion into n-type Si substrate was performed to form p-type base. This was done through an opening in a silicon dioxide mask and so there was a base mask for this step. (Note: Silicon dioxide can be used to mask boron diffusion into silicon.) Subsequently, phosphorus diffusion into

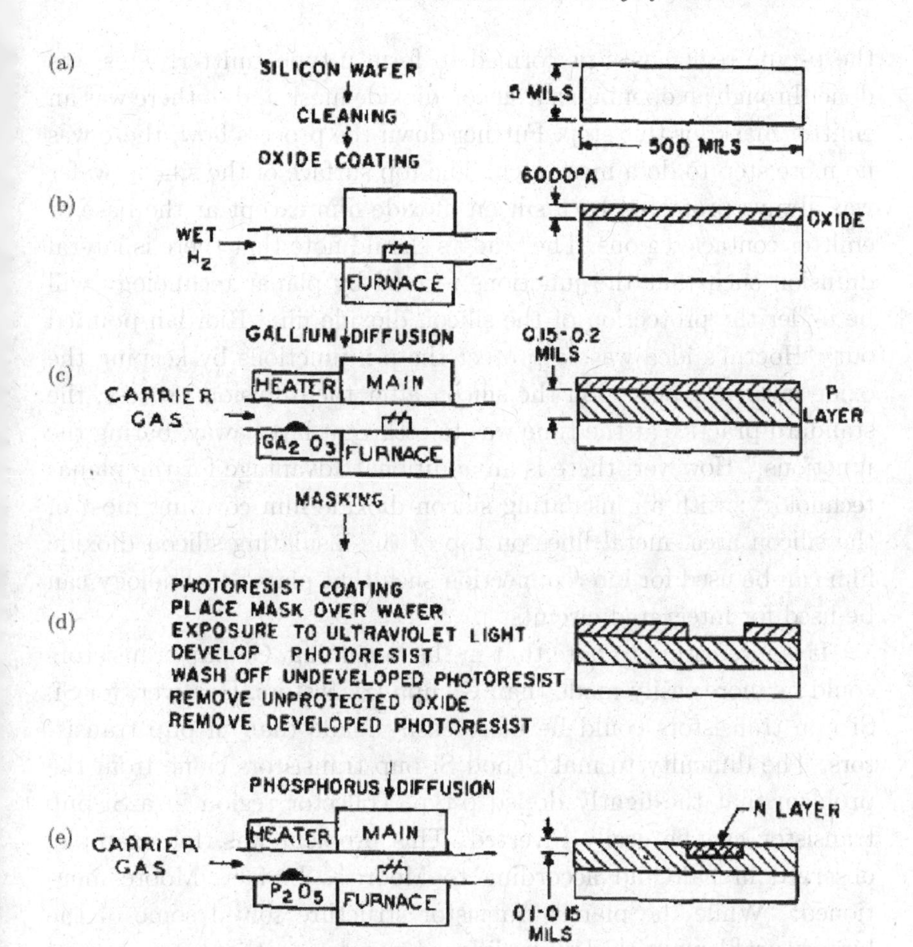

Fig. 1.4 Detailed diffusion and masking process according to Aschner *et al.* 1959.[48] A silicon dioxide film was first grown on n-type silicon and then gallium was diffused through the silicon dioxide film to form a p-type layer for the base of the npn transistor. A window was opened in the silicon dioxide film and then phosphorus was diffusion to form a localized n-type emitter. (Reprinted with permission from J. F. Aschner, C. A. Bittmann, W. F. J. Hare and J. J. Kleimack, "A double diffused silicon high-frequency switching Transistor produced by oxide masking techniques", *Journal of the Electrochemical Society,* vol. 106, no. 5 (May 1959) pp. 415–417. Copyright 1959 Electrochemical Society. Reproduced by permission of The Electrochemical Society.)

the p-type region was performed to form n-type emitter. This was done through an opening in a silicon dioxide mask and so there was an emitter mask for this step. Further down the process flow, there was no more step to do a mesa etch. The top surface of the silicon wafer was always protected by a silicon dioxide film except at the base or emitter contact regions. The readers should note that there is lateral diffusion such that the junctions formed by planar technology will be under the protection of the silicon dioxide film. Riordan pointed out: "Hoerni's idea was to protect the p-n junctions by keeping the oxide layer in place upon the silicon after the diffusion process; the standard practice at the time was to etch that layer away, baring the junctions." However, there is an additional advantage for the planar technology: with an insulating silicon dioxide film covering most of the silicon area, metal lines on top of the insulating silicon dioxide film can be used for interconnection such that planar technology can be used for integrated circuits.

It is interesting to note that in the beginning, Ge pnp transistors could be more easily made than Ge npn transistors. However, for Si, Si npn transistors could be more easily made than Si pnp transistors. The difficulty to make good Si pnp transistors came from the problem that the lightly doped p-type collector region in a Si pnp transistor can be easily inverted. This problem was, for example, observed in Fairchild according to Moore.[64] In fact, Moore mentioned: "While the planar transistor structure solved some of the biggest problems with double-diffused transistors, there were several that persisted. Particularly over the lightly doped collector region of high-voltage p-n-p planar transistors, an inversion layer sometimes developed, effectively extending the base region to the edge of the die. Inversion layers were not a new phenomenon. For example, early grown-junction transistors had problems with such layers developing on the surface of the base's shorting the emitter to the collector. In fact, the reason that Fairchild's first transistor had the base contact completely surrounding the emitter was to eliminate the possibility of such inversion layers." For discrete pnp transistors, this problem was usually solved by etching a mesa or by using a heavily doped p-type guard ring according to Finch and Haenichen.[65] Similar approach can also be found in other references.[66,67] For discrete pnp transistor, usually a heavily doped p-type guard ring is used to prevent possible inversion of the lightly doped p-type collector region; the mesa approach is usually not adopted probably because it is not compatible with "planar technology". However, the silicon bipolar integrated circuit is actually based on silicon npn transistors, maybe with some lateral pnp transistors. Naturally, Si npn transistors perform better than Si pnp transistors because electron mobility is larger than hole mobility, resulting in larger electron diffusion coefficient than hole diffusion coefficient through the Einstein relationship.

At this point, the author would also like to discuss some history about the development of photolithography. Johann Alois Senefelder (1771–1834) was an actor and playwright who became famous because of the invention of the printing technique of lithography in 1796. In 1955, Jules Andrus and Walter L. Bond at Bell Laboratories began the adaptation of existing photolithographic techniques already developed for making patterns on printed circuit boards to produce much finer patterns on silicon wafers to define patterns of a silicon dioxide layer as an impurity diffusion mask. After applying a photosensitive coating or "resist" on the layer and exposing the desired pattern on this coating through an optical mask, precise window areas were defined in the layer and opened by chemical etching where unexposed resist had been washed away. Impurities (n-type or p-type) were diffused through these silicon dioxide openings into the underlying silicon to create regions of n-type and p-type silicon needed in semiconductor devices. For this work, Andrus filed for a patent in 1957 and the patent was formally approved in 1964.[68] An article by Andrus and Bond was also published.[69] In 1957, Jay Lathrop and James Nall of the US Army's Diamond Ordnance Fuse Laboratories in Maryland filed for a patent on photolithography and the patent was formally approved in 1959.[70] A short paragraph about their work was published in 1958.[71] In 1959, Lathrop switched to Texas Instruments and worked for Jack Kilby while Nall switched to Fairchild Semiconductor. Following up on this pioneering work, Jay Last (1929–) and Robert Noyce (1927–1990) built one of the first "step-and-repeat" cameras at Fairchild Semiconductor in 1958 to make many identical transistors on a single silicon wafer using photolithography. In 1961, the David W. Mann division of GCA Corporation was the first firm to make commercial step and repeat mask reduction devices (photo-repeaters). This is now known as a "stepper". Subsequently "scanner" technology was developed. Previously, some scientists suggested that electron beam lithography or X-ray lithography may be necessary but photolithography remains an essential step in semiconductor manufacturing today, with feature sizes below 0.1 μm routinely generated with the help of deep UV technology.

1.5 History of Semiconductor Physics

In the previous paragraphs, the author has neglected those scientists responsible for theoretical development and basic physics. Here the author would like to mention some other prominent scientists directly or indirectly involved in the semiconductor revolution. Georg Simon Ohm (1789–1854) was a German scientist well known because of the Ohm's Law. Ohm's law, which states that the electric current is proportional to the potential difference, was first discovered by a British scientist Henry Cavendish (1731–1810). However, Cavendish did not publish his discovery but Ohm published it such that it subsequently came to bear his name. The law appeared in Ohm's famous book *Die galvanische Kette, mathematisch bearbeitet* (The Galvanic Circuit Investigated Mathematically) published in 1827 in which he gave his complete theory of electricity. Nowadays, Ohm's Law is considered something straightforward but in Ohm's days, his work was actually received with little enthusiasm in the beginning. However, his work was eventually recognized by the Royal Society of London. Ohm received the Copley Medal in 1841. Augustus Matthiessen (1831–1870) was a British physicist and chemist who has been indirectly present in semiconductor books because of the Matthiessen's rule for carrier mobility. The Matthiessen's rule for carrier mobility probably originated from Augustus Matthiessen's study of electrical conduction of metals and alloys.[72–74] In Matthiessen's time, the concept of "mobility" was not established yet. The modern form of

Matthiessen's rule for electron mobility or hole mobility is actually an extension of Matthiessen's work in the 19th century by subsequent scientists. It is sad that Matthiessen was a tragical scientist who committed suicide in 1870 under severe nervous strain. Edwin Herbert Hall (1855–1938) was an American physicist who discovered the Hall Effect.[75] The above discoveries were made before the discovery of the electron. Eventually, a British physicist Sir Joseph John Thomson (1856–1940) discovered the electron in 1897.[76] However, it was an Irish physicist George Johnstone Stoney (1826–1911) who coined the name "electron". Furthermore, it was Paul Drude (1863–1906), a prominent German scientist, who proposed the first theory of electronic conduction,[77] resulting in Drude's model. Busch, a pioneer of semiconductor research, pointed out that, besides Paul Drude, Eduard Riecke was a much less famous scientist who had also worked on electronic conduction.[1] However, like Matthiessen, Paul Drude was another tragical scientist who committed suicide in 1906. Drude's model was later refined by Arnold Sommerfeld (1868–1951) and Hans Bethe (1906–2005).[78] Erwin Schrodinger (1887–1961) was born in Austria. Once he worked in Germany. Eventually he immigrated to Ireland. He is famous for his Schrodinger equations which are important for solving problems in quantum electronics. Felix Bloch (1905–1983) was a Swiss Jew who left Germany and emigrated to the United States when Adolf Hitler came to power in Germany. (Note: There was another Felix Bloch who was born in 1935 and involved in an espionage case.) He was born and educated in Zurich, Switzerland. However, he went to the University of Leipzig, Germany in 1928 for his PhD and Werner Heisenberg (1901–1976), who was a prominent German scientist and famous for the uncertainty principle, was his PhD advisor. At that time, Schrodinger equations were already known. His PhD dissertation provided the theory of electrons in crystal lattices which is the basis for the quantum theory of electrical conduction. According to Bloch,[79] the electrons can move without scattering if the crystal lattice is perfect and there is no lattice vibration. The original Bloch paper was in German. Subsequently, he published an article in English regarding the history of his 1928 work.[80] The significance of Bloch's work for this book is that electrons in silicon can be modeled as Bloch waves. The E-k diagram of silicon shows the energy E, which is a scalar, plotted against the wavevector **k**. E and the 3-D vector **k** are two important parameters for 3-D Bloch waves. The direction of the 3-D vector **k** represents the direction of the movement of a 3-D wave. Sir Alan Herries Wilson (1906–1995) was a British physicist responsible for the modern band theory. In 1930, he recognized the difference between conductors and insulators; conductors have only partially-filled upper energy bands so that electrons in this band can acquire kinetic energy; the upper energy band is filled in an insulator. In a semiconductor, the presence of impurities contribute electrons to the empty upper energy band. Whereas Bloch modeled electrons as waves, Wilson explained the difference between metals, semiconductors and insulators using band theory.[81–83] The valence band and the conduction band of silicon are quite frequently represented by an E-k diagram where the energy E, which is a scalar, is plotted against the wavevector **k**. Cahn published a short article in 2005 to praise Wilson.[84] He pointed out:

Until the end of the 1930s, most physicists looked down their noses at semiconductors and kept clear of them. The man who changed all this was Alan Herries Wilson, a theoretical physicist in Cambridge, who as a young man spent a sabbatical with Heisenberg in Leipzig and applied the brand new field of quantum mechanics to issues of electrical conduction, first in metals and then in semiconductors, as reported in two Royal Society papers in 1930 and 1931. When he returned to Cambridge, Wilson urged that attention be paid to germanium but, as he expressed it long afterward, "the silence was deafening" in response. He was told that devoting attention to semiconductors, those messy entities, was likely to blight his career among physicists. He ignored these warnings and in 1939 brought out his famous book, *Semiconductors and Metals*, which explained semiconductor properties, including the much-doubted phenomenon of intrinsic semiconductivity, in terms of electronic energy bands. His academic career seems indeed to have been blighted, because despite his great intellectual distinction, he was not promoted in Cambridge (he remained an assistant professor year after year). At the end of World War II, he abandoned his university functions and embarked on a notably successful career as a captain of industry, culminating in his post of chief executive of a leading British pharmaceutical company; he kept clear of electronics! In due course he became Sir Alan Wilson.

The British pharmaceutical company mentioned above was GlaxoSmithKline (GSK).[85] In his Nobel lecture delivered in 1956, John Bardeen, who invented the point contact transistor, quoted Wilson's work as his first reference. Thus Wilson's contribution to the basic understanding of semiconductors should not be forgotten.

Methods were developed to compute the E-k diagram for electrons in silicon and germanium, for example, by Herman.[86] Because of periodicity in k space, it is enough to show the E-k diagram in the first Brillouin zone. The concept of Brillouin zone was developed by Leon Nicolas Brillouin (1889–1969), who was a French physicist. Besides the E-k diagram for electrons in silicon, there exists another E-k diagram for phonons in silicon. Because of periodicity in k space, it is enough to show the E-k diagram for phonons in the first Brillouin zone. The quantum of energy in an elastic wave in a crystal was named "phonon" in direct analogy to the photon. According to Walker and Slack, the name probably came from the Soviet scientists Igor Yevgenyevich Tamm (1895–1971) and Yakov Ilich Frenkel (1894–1952).[87] For semiconductor people, Tamm is famous for Tamm surface states. (Note: I. Tamm (1932). *Phys. Z. Soviet Union* **1**: 733.) Frenkel is famous for the Poole-Frenkel effect for modeling the leakage current of insulators. Bertram Neville Brockhouse (1918–2003) was a Canadian physicist who developed the neutron scattering technique used to measure the E-k diagram for phonons in silicon.[88–90] He shared the 1994 Nobel Prize in Physics with another American scientist, Clifford Glenwood Shull (1915–2001), who also worked on neutron scattering. Brockhouse published a review paper in 1995.[91] The optical phonon energy of 63 meV is something semiconductor people have to memorize.

Maxwell-Boltzmann statistics was named after James Clerk Maxwell (1831–1879) and also after Ludwig Boltzmann (1844–1906). Maxwell is famous for Maxwell's equations in electromagnetism. Boltzmann's constant has been frequently used in semiconductor text books. It is a great tragedy that he hanged himself in 1906.[92] Although Boltzmann first linked entropy and probability in 1877, it seems the relation was never expressed with a specific constant until Max Planck (1858–1947) first introduced k, and gave an accurate value for it (1.346×10^{-23} J/K, about 2.5% lower than today's figure), in his derivation of the law of black body radiation in 1900–1901. Before 1900, equations involving Boltzmann factors were not written using the energies per molecule and Boltzmann's constant, but rather using a form of the molar gas constant R divided by the Avogadro's number. Thus the Boltzmann's constant was not created by Boltzmann but was named after him by Max Planck (1858–1947). Fermi-Dirac statistics was developed by Enrico Fermi (1901–1954)[93] and also independently by Paul Adrien Maurice Dirac (1902–1984).[94] Fermi was an Italian physicist who later became a US citizen.[95] Dirac was a UK physicist who contributed significantly to the development of quantum mechanics.[96] In addition, the Einstein's relationship was named after Albert Einstein (1879–1955). Einstein did not work on semiconductors. However, he once worked on Brownian motion.[97–99] The Einstein's relationship for electrons and holes in semiconductors was an extension of Einstein's work on Brownian motion.

1.6 History of Semiconductor Crystal Growth Technology

Besides the device physics and technology, the basic semiconductor material is also important. Thus crystal growth is also an important concern for semiconductor people. Jan Czochralski (1885–1953) was born in Exin, German Empire which existed from 1871 to 1918. He was an ethnic Pole. Previously, Poland suffered from the First Partition of Poland in 1772 by Russia, Prussia and Austria; Poland suffered from the Second Partition of Poland in 1793 by Russia and Prussia; finally Poland suffered from the Third Partition of Poland in 1795 by Russia, Prussia and Austria. In this way, when Jan Czochralski was born in 1885, Poland did not exist as an independent country. Around 1900, he moved to the German capital Berlin, where he worked at a pharmacy. He was educated at

Charlottenburg Polytechnic in Berlin, where he specialized in metal chemistry. Czochralski began working as an engineer for Allgemeine Elektrizitäts Gesellschaft (AEG) in 1907. After War World I, the German Empire became the Weimar Republic and the nation of Poland was resurrected. Anyhow, his birthplace is now known as Kcynia, Poland. In 2004, Kcynia showed a population of only about 4000–5000 and so this is actually a small Polish village. Thus he is now usually considered a Polish scientist. During World War I, he discovered the Czochralski method in 1916, when he accidentally dipped his pen into a crucible of molten tin rather than his inkwell. He immediately pulled his pen out to discover that a thin thread of solidified metal was hanging from the nib of his pen. The nib was replaced by a capillary and Czochralski verified that the crystallized metal was a single crystal. The experiments of Czochralski produced single crystals that were a mm in diameter and up to 150 cm long. Czochralski published a paper on his discovery in 1918 in the Zeitschrift für Physikalische Chemie, a German chemistry journal, with the title "Ein neues Verfahren zur Messung der Kristallisationsgeschwindigkeit der Metalle". The German title can be translated into English as "A new method for the measurement of the crystallization rate of metals". In fact, the method was at that time used for measuring the crystallization rate of metals such as tin, zinc and lead.[100]

During his stay in Germany, Czochralski wrote several papers, patents and books and was a member of several scientific societies. With his German friends he founded in 1919 German Society for Metals Science (Deutsche Gesellschaft für Metallkunde) and he was its president in 1925. After World War I, a new independent Republic of Poland was created near the end of 1918 with Warsaw as capital. In 1929, the President of Poland, Ignacy Moscicki (1867–1946, a professor of chemistry who served as the President of Poland from 1926 to 1939) invited him to Poland and he received the position of a professor in the Faculty of Chemistry of the Warsaw University of Technology. At the same time he obtained honorary doctorates. Professor J. Czochralski worked for about 30 years in Germany and from 1928 in Warsaw, Poland. He worked in the Warsaw University of Technology as a professor. Nazi Germany invaded Poland in 1939 and started World War II. After World War II, he was stripped of his professorship due to his involvement with Germany during the war, although he was later cleared of any wrongdoing by a Polish law court. He returned to his native town of Kcynia where he ran a small cosmetics and household chemicals firm until his death in 1953. Some people claimed that Czochralski had been helping the Polish resistance during the German occupation. Some people claimed that he had been a German collaborator. However, nowadays, he is usually considered a hero in Polish science and technology.[101] A website www.janczochralski.com has been dedicated to the memory of Jan Czochralski.

In 1950, Gordon K. Teal and John B. Little, who were Americans working in the Bell Laboratories used this method to grow germanium single crystals.[102,103] Later a similar approach was used to grow silicon single crystals. Subsequently, Teal left the Bell Laboratories to join TI. In TI, Teal developed the first workable silicon transistors as discussed above.

When Czochralski worked on crystal growth by drawing, he was probably a German citizen working in a German company in the German Empire and he published in German in a German journal. In some way, it can be considered a German invention. However, Czochralski can be considered an ethnic Pole. In fact, he later became a professor in Poland. As mentioned above, the method became mature because of two Americans working in the Bell Laboratories. Thus, to be fair to all parties, the invention can be considered a German/Polish/US invention. The Czochralski technique is currently the principal technology to grow large silicon single crystal. Czochralski silicon wafers are usually contaminated by oxygen. However, oxygen contamination is not so bad. It affects the mechanical strength of the silicon. It can also help to getter impurities. The Czochralski silicon single-crystal growth technology is still intensively studied.

As discussed earlier in this chapter, how impurities in dissolved germanium or silicon can end in the frozen germanium or silicon have been studied for many years. Various references for impurity distribution or segregation include.[105–109] Silicon crystals grown by the Czochralski method tends to be contaminated by oxygen. Silicon may be contaminated by various impurities during crystal growth or subsequent processing. "Gettering" of impurities can be important. George Bemski (Bell Laboratories) was one of the pioneers working in this research area. Various references for gettering include.[110–118] In the book, *History of Semiconductor Engineering* by Lojek,[119] there is a discussion on how npn and pnp transistors are fabricated. There is a step of depositing "nickel" on the back side of the silicon wafer, a step of heating the silicon wafer to high temperature with the nickel film present on the back side and a step of removing the nickel film. Lojek did not explain the function of these steps. It was Moore[64] who mentioned that the nickel film was used to getter impurities from the silicon wafer. However, it should be noted nickel can getter metallic impurities but nickel by itself is also a metallic impurity. Thus it is possible to use nickel to getter impurities, resulting in higher minority carrier lifetime and thus higher current, or to use nickel as an impurity to degrade minority carrier lifetime for better switching speed. It just depends on the exact processing conditions.

1.7 Semiconductor Science and Technology in the 21st Century

In conclusion, selenium may be the first elemental semiconductor studied and used but its significance has dropped very significantly. However, research papers on selenium can still be found in the 21st century.[120,121] PbS is still used for infrared applications. By the way, one of the oldest semiconductor with great industrial significance was cuprous oxide (Cu_2O). It is a semiconductor with a bandgap of about 2 eV. Cu/CuO_2 rectifiers once were very commonly used in industry. Lars Olai Grondahl published a review on cuprous oxide in 1931[122] while Brittain, one of the three inventors of transistor, also published a review on cuprous oxide in 1951.[123] Cu/CuO_2 rectifiers are no longer popular. However, research papers on cuprous oxide are still published in the second half of the 20th century[124] and even in the 21st century.[125,126] Germanium was once the main semiconductor used for transistors. Professor Karl Lark-Horovitz, Department of Physics of Purdue University, put in a great effort to study it during

World War II.[127,128] However, it was Bell Laboratories scientists who managed to make the first germanium transistor. It was subsequently replaced by silicon. Silicon grown by the Czochralski technique is currently the dominant semiconductor material studied and used in the semiconductor industry. An alloy of silicon and germanium became important in the semiconductor industry since the 1990's. Recently, there is a chance that germanium may come back as an important semiconductor used in CMOS integrated circuits.[129] In addition, we shall see later in this book that some research groups have been seriously investigating the application of III-V semiconductors to mainstream CMOS integrated circuits. Various references have also been consulted to write this chapter.[130–136] At this point, the author would also like to emphasize the indirect contribution of the physicists to semiconductor physics and technology. For example, Sir Neville Francis Mott (1905–1996) has been famous for his Mott's rule regarding the barrier height of metal-semiconductor junctions[137]: the barrier height is given by the difference between the metal work function and the semiconductor electron affinity. However, Mott's rule has been found to be insufficient to guide the technology of making Ohmic contacts on semiconductors. In 1956, Kroger *et al.* pointed out that a strongly doped surface layer and quantum mechanical tunneling are important for making good Ohmic contacts.[138] Albert Yu followed up on this work in 1970.[139,140] As shown in Fig. 3, Aschener *et al.* used Au-Sb for the n-type emitter contact and Al for the p-type contact. Gordon Moore, one of the three founder of Intel, pointed out that a single metal can be used to contact the p-type base and the n-type emitter.[64] Without the understanding of quantum mechanics, this may not be easily explained. Al is p-type dopant for Si such that Al can form a heavily doped layer in p-type Si after heating, resulting in a good Ohmic contact. As long as the n-type emitter is heavily doped, Al can also form a good Ohmic contact. With this in mind, it is easy to understand that a single silicide can be used to make good Ohmic contact for both n-channel and p-channel MOS transistors as long as the drain/source regions are heavily doped. For example, state-of-the-art CMOS technology uses cobalt silicide or nickel silicide to make contacts to the gate, drain and source for both n-channel and p-channel devices. Recently, the research on Ohmic contacts for Ge-based CMOS is a hot research topic.[141] As we shall see later in this book, quantum mechanical tunnelling is also a source of major current leakage in CMOS, and results in the substantial power drain and heating effects that plague high-speed and mobile technology. Merzbacher gave a history of the early days of the theory of quantum mechanical tunneling.[142] Sir Ralph Howard Fowler (1889–1944) and Lothar Wolfgang Nordheim (1899–1985) were probably the first scientists to apply quantum mechanics to electron tunneling.[143] Thus those scientists who developed the theories of quantum mechanics also indirectly contribute to semiconductor science and technology. At this point, the reader may wonder why the author spent so much time on subjects apparently not related to CMOS technology. As discussed above, Ohmic contacts are important even in state-of-the-art CMOS technology and Ohmic contact physics and technology cannot be understood without the concept of quantum mechanical tunneling. As discussed above, in the early days of semiconductor technology, Ohmic contact to p-type Ge or Si and Ohmic contact to n-type Ge or Si employ different metals. In state-of-art CMOS technology, nickel silicide (NiSi) is used for Ohmic contact to both p-type and n-type Si; as long as the p-type and n-type Si is heavily doped enough, a single type of metal can be used as Ohmic contact according to quantum mechanical tunneling theory. In addition, as pointed out by E. O. Johnson,[144] the MOS transistor is actually a bipolar transistor in disguise; there is an npn parasitic transistor in an n-channel enhancement-mode MOS transistor while there is a pnp parasitic transistor in a p-channel enhancement-mode MOS transistor. The phenomenon of punchthrough is observed in both bipolar transistors[145,146] and MOS transistors.[147,148] Thus the understanding of bipolar transistor physics and technology is also important for the understanding of CMOS physics and technology. By the way, the understanding of MOS physics also has an important impact on bipolar transistor physics and technology; references include.[149–157] As discussed above, the lightly doped p-type collector of discrete pnp transistors may invert; a commonly known solution is to add a p^+ guard ring. Finally, the discussion of the Bloch Theorem, Brillouin zone and band structure of semiconductors can be seen as important later in this book; for example, the author will later point out that rotation by 45° from the conventional $\langle 110 \rangle$ direction to the $\langle 100 \rangle$ direction can lead to better p-channel transistors and this can be predicted by understanding of the valence band structure of silicon.

References

[1] G. Busch, "Early history of the physics and chemistry of semiconductors — from doubts to fact in a hundred years", *European Journal of Physics*, vol. 10 (1989), pp. 254–264.

[2] A. Volta, "Del modo di render sensibilissima la piu debole elettricita sia naturale, sia artificiale", *Philosophical Transactions of the Royal Society of London*, vol. 72 (1782), pp. 237–280. The accompanying English translation appeared as: A. Volta, "Of the method of rendering very sensible the weakest natural or artificial electricity", *Philosophical Transactions of the Royal Society of London*, vol. 72 (1782), pp. vii–xxxiii.

[3] H. Davy, "Farther researches on the magnetic phenomena produced by electricity; with some new experiments on the properties of electrified bodies in their relations to conducting powers and temperature", *Philosophical Transactions of the Royal Society of London*, vol. 111 (1821), pp. 425–439.

[4] M. Faraday, "Experimental researches in electricity, fourth series", *Philosophical Transactions of the Royal Society of London*, vol. 123 (1833), pp. 507–522.

[5] G. A. Martínez-Castañón, M. G. Sánchez-Loredo, H. J. Dorantes, J. R. Martínez-Mendoza, G. Ortega-Zarzosa and F. Ruiz, "Characterization of silver sulfide nanoparticles synthesized by a simple precipitation method", *Materials Letters*, vol. 59, no. 4 (Feb. 2005), pp. 529–534.

[6] W. S. Lau, P. Yang, V. Ho, L. F. Toh, Y. Liu, S. Y. Siah and L. Chan, "An explanation of the dependence of the effective saturation velocity on gate voltage in sub-0.1 μ m metal-oxide-semiconductor transistors by quasi-ballistic transport theory", *Microelectronics Reliability*, vol. 48, no. 10 (Oct. 2008), pp. 1641–1648.

[7] L. Vadasz and A. S. Grove, "Temperature dependence of MOS transistor characteristics below saturation", *IEEE Trans. Electron Dev.*, vol. 13, no. 12 (Dec. 1966), pp. 863–866.

[8] I. M. Filanovsky and A. Allam. "Mutual compensation of mobility and threshold voltage temperature effects with applications in CMOS circuits", *IEEE Trans. Circuits and Systems- I: Fundamentals Theory and Application*, vol. 48, no. 7 (July 2001), pp. 876–884.

[9] K. Hisamitsu, H. Ueno, M. Tanaka, D. Kitamaru, M. Miura-Mattausch, H. J. Mattausch, S. Kumashiro, T. Yamaguchi, K. Yamashita and N. Nakayama, "Temperature-independence-point properties for 0.1 μm-scale pocket-implant technologies and the impact on circuit

design", *Proceedings of the 2003 Asia and South Pacific Design Automation Conference* (ASP-DAC 2003, IEEE), pp. 179–183.

[10] L. W. Yan and M. Chan, "Effect of technology scaling on temperature independent point (TIP) in MOS transistors", 8th *International Conference on Solid-State and Integrated Circuit Technology* (ICSICT'06, IEEE), Shanghai, China 2006, pp. 203–205.

[11] P. Yang, W. S. Lau, S. W. Lai, V. L. Lo, S. Y. Siah and L. Chan, "Selection of gate length and gate bias to make nanoscale metal-oxide-semiconductor transistors less sensitive to both statistical gate length variation and temperature variation", *Solid-State Electron.*, vol. 54, no. 11 (Nov. 2010), pp. 1304–1311.

[12] A. E. Becquerel, " Mémoire sur les effets électriques produits sous l'influence des rayons solaires", *Comptes Rendus*, vol. 9 (1839), pp. 561–567.

[13] W. W. Anderson and Y. G. Chai, "Becquerel effect solar effect", *Energy Conversion*, vol. 15, nos. 3–4 (1976), pp. 85–94.

[14] K. Kalyanasundaram, "Photoelectrochemical cell studies with semiconductor electrodes — A classified bibliography (1975–1983)", *Solar Cells*, vol. 15 (1985), pp. 93–156.

[15] W. Smith, "Effect of light on selenium during the passage of an electric current", *Nature*, 20 February 1873, p. 303.

[16a] W. G. Adams and R. E. Day, "The action of light on selenium", *Proceedings of the Royal Society of London*, vol. 25 (1876–1877), pp. 113–117.

[16b] W. G. Adams and R. E. Day, "The action of light on selenium", *Philosophical Transactions of the Royal Society of London*, vol. 167 (1877), pp. 313–349.

[17] C. Fritts, "A new form of selenium cell", *American Journal of Science*, vol. 26 (1883), pp. 465–472.

[18] K. Lehovec, "The photo-voltaic effect", *Phys. Rev.*, vol. 74, no. 4 (August 15, 1948), pp. 463–471.

[19] D. Owen, *Copies in Seconds: Chester Carlson and the Birth of the Xerox Machine*, Simon & Schuster, New York, 2004, pp. 96–100.

[20] F. Braun, "Uber die stromleitung durch schwefelmetalle", Annalen der Physik und Chemie, vol. 153 (1874) pp. 556–563. The German version of this paper has been reprinted in the book "Electronic structure of metal-semiconductor contacts", edited by W. Monch, Kluwer Academic Publishers, Dordrecht/Boston/London, 1990, pp. 556–563. In addition, it has been translated into English and reprinted in the book *Semiconductor Devices: Pioneering Papers*, edited by S. M. Sze, World Scientific, Singapore, 1991: F. Braun, "On current conduction through metallic sulfides", pp. 377–380.

[21] H. J. Lian, A. Yang, M. L. W. Thewalt, R. Lauck and M. Cardona, "Effects of sulfur isotopic composition on the band gap of PbS", *Phys. Rev. B*, vol. 73, no. 23, pp. 233202–1 to 233202–4 (15 June 2006).

[22] W. Schottky, "Halbleitertheorie der sperrschicht", Naturwissenschaften, vol. 26 (1938) p. 843. This paper has been translated into English and reprinted in the book *Semiconductor Devices: Pioneering Papers*, edited

by S. M. Sze, World Scientific, Singapore, 1991: W. Schottky, "Semiconductor theory of the blocking layer", p. 381. This 1938 paper is less than one page. Subsequently, Schottky published two much longer papers: W. Schottky, "Zur halbleitertheorie der sperrschicht und spitzengleichrichter", *Zeitschrift fur Physik*, vol. 113, no. 5–6 (1939) pp. 367–414. The title translated from German into English is "Semiconductor theory of blocking layer and point rectifier". W. Schottky, "Vereinfachte und erweiterte theorie der randschicht-gleichrichter", *Zeitschrift fur Physik*, vol. 118, no. 9–10 (1942) pp. 539–592. The title translated from German into English is "Simplified and extended theory of boundary layer rectifier".

[23] J. C. Bose, "Detector for electrical disturbances", US Patent 755,840, filed in 1901 and awarded in 1904. (Note: This is the patent for galena detector, etc.)

[24] G. W. Pickard, "Means for receiving intelligence communicated by electric waves", US Patent 836,531, filed in 1906 and awarded in 1906.

[25] J. A. Fleming, "Instrument for converting alternating electric currents into continuous current", US Patent 803,684, filed in 1905 and awarded in 1905. (Note: This is the patent for vacuum tube diode.)

[26] L. de Forest, "Space telegraphy", US Patent 879,532, filed in 1907 and awarded in 1908. (Note: This is the patent for vacuum tube triode.)

[27] G. Holst and B. D. H. Tellegen, "Means for amplifying electric oscillations", US Patent 1,945,040, filed in 1927 and awarded in 1934. (Note: This is the patent for vacuum tube pentode.)

[28] R. Buderi, "The invention that changed the world: How a small group of radar pioneers won the second world war and launched a technical revolution", Touchstone, New York, 1996, pp. 308–333 (Chapter 15).

[29] J. Orton, *The Story of Semiconductors*, Oxford University Press, Oxford, UK, 2004, pp. 38–42.

[30] F. B. Llewellyn, "Vacuum tube electronics at ultra-high frequencies" *Proc. IRE*, vol. 21, no. 11 (Nov. 1933), pp. 1532–1573.

[31] F. Seitz, "Research on silicon and germanium in World War II", *Physics Today*, vol. 48, no. 1 (Jan. 1995), pp. 22–27.

[32] J. H. Scaff and R. S. Ohl, "Development of silicon crystal rectifiers for microwave radar receivers", *Bell System Technical Journal*, vol. 26, no. 1 (Jan. 1947), pp. 1–30. L. W. Morrison, Jr., "The radar receiver", Bell System Technical Journal, vol. 26, no. 4 (Oct. 1947), pp. 694–817.

[33] M. Riordan and L. Hoddeson, "The origins of the pn junction", *IEEE Spectrum*, vol. 34, no. 6 (June 1997), pp. 46–51.

[34] M. Riordan and L. Hoddeson, *Crystal Fire: The Invention of the Transistor and the Birth of the Information Age*, Norton, New York, 1997.

[35] R. S. Ohl, "Light-sensitive electrical device", US Patent 2,402,662, filed in 1941 and awarded in 1946.

[36] R. S. Ohl, "Alternating current rectifier", US Patent 2,402,661, filed in 1941 and awarded in 1946.

[37] J. H. Scaff, H. C. Theuerer and E. E. Schumacher, "P-type and n-type silicon and the formation of the photovoltaic barriers in silicon ingots", *Trans. A.I.M.E.*, vol. 185, pp. 383–388, June 1949.

[38] J. R. Woodyard, "Nonlinear circuit device utilizing germanium", US Patent 2,530,110, filed in 1944 and awarded in 1950.

[39] J. Bardeen and W. H. Brattain, "The transistor: a semi-conductor triode", *Phys. Rev.*, vol. 74, no. 2, pp. 230–231 (15th July, 1948).

[40] J. Bardeen and W. H. Brattain, "Physical principle involved in transistor action",*Phys. Rev.*, vol. 75, no. 8, pp. 1208–1225 (15th April, 1949).

[41] J. Bardeen and W. H. Brattain, "Three-electrode circuit element utilizing semiconductor materials", US Patent 2,524,035, filed in 1948 and awarded in 1950.

[42] W. Shockley, M. Sparks and G. K. Teal, "p-n junction transistors", *Phys. Rev.*, vol. 83, no. 1, pp. 151–162 (1st July 1951).

[43] W. Shockley, "Circuit element utilizing semiconductor material", US Patent 2,569,347, filed in 1948 and awarded in 1951.

[44] A. Van Dormael, "The French transistor", *Proceedings of the 2004 IEEE Conference on the History of Electronics, Bletchley Park*, England (June 2004). (http://www.ieee.org/organizations/history_center/Che2004/VanDormael.pdf)

[45] M. Riordan, "How Europe missed the transistor", *IEEE Spectrum*, vol. 42, no. 11 (November 2005), pp. 52–57.

[46] M. Riordan, "The lost history of the transistor", *IEEE Spectrum*, vol. 41, no. 5 (May 2004), pp. 44–49.

[47] M. Tanenbaum and D. E. Thomas, "Diffused emitter and base silicon transistors," *Bell System Technical Journal*, vol. 35, no. 1 (January 1956), pp. 1–15.

[48] J. F. Aschner, C. A. Bittmann, W. F. J. Hare and J. J. Kleimack, "A double diffused silicon high-frequency switching Transistor produced by oxide masking techniques", *Journal of the Electrochemical Society*, vol. 106, no. 5 (May 1959) pp. 415–417.

[49] H. C. Theuerer, J. J. Kleimack H. H. Loar and H. Christensen, "Epitaxial diffused transistors" *Proc. IRE*, vol. 48, no. 9 (Sep. 1960), pp. 1642–1643.

[50] R. R. Roup and J. S. Kilby, "Electrical circuit elements", US Patent 2,841,508, filed in 1955 and awarded in 1958.

[51] J. S. Kilby, "Invention of the integrated circuit", *IEEE Trans. Electron Dev.*, vol. 23, no. 7 (July 1976), pp. 648–654.

[52] J. S. Kilby, "The integrated circuit's early history", *Proc. IEEE*, vol. 88, no. 1 (Jan. 2000), pp. 109–111.

[53] R. N. Noyce, "Semiconductor device-and-lead structure", US Patent 2,981,877, filed in 1959 and awarded in 1961.

[54] R. M. Warner, "Microelectronics: Its unusual origin and personality", *IEEE Trans. Electron Dev.*, vol. 48, no. 11 (November 2001), pp. 2457–2467.

[55] A. N. Saxena, *Invention of Integrated Circuits: Untold Important Facts*, World Scientific, Singapore, 2009, pp. 1–523.
[56] K. Lehovec, "Multiple semiconductor assembly", US Patent 3,029,366, filed in 1959 and awarded in 1962.
[57] K. Lehovec, "Invention of p-n junction isolation in integrated circuits", *IEEE Trans. Electron Dev.*, vol. 25, no. 4 (April 1978), pp. 495–496.
[58] J. A. Hoerni, "Method of manufacturing semiconductor devices", US Patent 3,025,589, filed in 1959 and awarded in 1962.
[59] J. A. Hoerni, "Semiconductor device", US Patent 3,064,167, filed in 1959 and awarded in 1962.
[60] J. A. Hoerni, "Planar silicon diodes and transistors", *IRE Trans. Electron Dev.*, vol. 8, no. 2 (March 1961), p. 178.
[61] M. Riordan, "The silicon dioxide solution: How physicist Jean Hoerni built the bridge from the transistor to the integrated circuit", *IEEE Spectrum*, vol. 44, no. 12 (December 2007), pp. 51–56.
[62] L. Derick and C. J. Frosch, "Oxidation of semiconductive surfaces for controlled diffusion", US Patent 2,802,760, filed in 1956 and awarded in 1957.
[63] C. J. Frosch and L. Derick, "Surface protection and selective masking during diffusion in silicon", *J. Electrochem. Soc.*, vol. 104, no. 9 (September 1957), pp. 547–552.
[64] G. M. Moore, "The role of Fairchild in silicon technology", *Proc. IEEE*, vol. 86 no. 1 (Jan. 1998), pp. 53–62.
[65] J. R. Finch and J. C. Haenichen, "Annular — A new semiconductor device structure", *IEDM Tech. Dig.*, p. 98, 1963.
[66] E. D. Metz, "Silicon transistor failure mechanisms caused by surface charge separations", *Second Annual Symposium on the Physics of Failure in Electronics*, pp. 163–172, 1963.
[67] D. Zerrweck and N. Peters, "*Technologie der PNP-planartransistoren*", 1978, pp. 1–58.
[68] J. Andrus, "Fabrication of semiconductor devices", US Patent 3,122,817, filed in 1957 and awarded in 1964.
[69] J. Andrus and W. L. Bond, "Photoengraving in transistor fabrication", in *Transistor Technology*, vol. III, edited by F. J. Bioni, D. Van Nostrand, Princeton, N. J., 1958, pp. 151–162.
[70] J. W. Lathrop and J. R. Nall, "Semiconductor construction", US Patent 2,890,395, filed in 1957 and awarded in 1959.
[71] J. R. Nall and J. W. Lathrop, "Photolithographic fabrication techniques for transistors which are an integral part of a printed circuit", *IRE Trans. Electron Dev.*, vol. 5, no. 2 (April 1958), p. 117.
[72] A. Matthiessen, "On the electric conducting power of the metals", *Philosophical Transactions of the Royal Society of London*, vol. 148 (1858), pp. 383–387.
[73] A. Matthiessen and C. Vogt, "On the influence of temperature on the electric conductive-power of thallium and iron", *Philosophical Transactions of the Royal Society of London*, vol. 153 (1863), pp. 369–383.

[74] A. Matthiessen and C. Vogt, "On the influence of temperature on the electric conductive-power of alloys", *Philosophical Transactions of the Royal Society of London*, vol. 154 (1864), pp. 167–200.
[75] E. H. Hall, "On a new action of the magnet on electric currents", *American Journal of Mathematics*, vol. 2 (1879), pp. 287–292.
[76] J.J. Thomson, "Cathode rays", *Philosophical Magazine*, vol. 44 (1897), p. 293.
[77] P. Drude, "Zur elektronentheorie der metalle", *Annalen der Physik*, vol. 3, no. 3 (Nov. 1900), pp. 369–402. (Note: The title translated from German into English is "Electron theory of metals".)
[78] A. Sommerfeld and H. Bethe, "Elektronentheorie der Metalle", in *Handbuch der Physik*, Springer, Berlin, 1933, edited by H. Geiger and K. Scheel, vol. 24, Part 2, pp. 333–622. (Note: This nearly 300-page chapter was later published as a separate book, "Elektronentheorie der Metalle", Springer, 1967.)
[79] F. Bloch, "Quantum mechanics of electrons in crystal lattices (title translated into English)", *Zeitschrift fur Physik*, vol. 52, no. 7–8 (1928), pp. 555–600.
[80] F. Bloch, "Memories of electrons in crystals", *Proceedings of the Royal Society A*, vol. 371, no. 1744 (Jun. 10, 1980), pp. 24–27.
[81] A. H. Wilson, "The theory of metals I", *Proceedings of the Royal Society A*, vol. 138, no. 836 (Dec. 1, 1932), pp. 594–606.
[82] A. H. Wilson, "The theory of electronic semi-conductors", *Proceedings of the Royal Society A*, vol. 133, no. 822 (Oct. 1, 1931), pp. 458–491.
[83] A. H. Wilson, "The theory of electronic semi-conductors II", *Proceedings of the Royal Society A*, vol. 134, no. 823 (Nov. 3, 1931), pp. 277–287.
[84] R. W. Cahn, "Silicon: a child of revolution", The Electrochemical Society Interface, Spring 2005, pp. 15–16.
[85] E. H. Sondheimer, "Sir Alan Herries Wilson 2 July 1906–30 September 1995", Biographical Memoirs of Fellows of the Royal Society, vol. 45 (Nov. 1999), pp. 549–562.
[86] F. Herman, "The electronic energy band structure of silicon and germanium", *Proc. IRE*, vol. 43, no. 12 (Dec. 1955), pp. 1703–1732.
[87] C. T. Walker and G. A. Slack, "Who named the -ons", *American Journal of Physics*, vol. 38, no. 12 (Dec. 1970), pp. 1380–1389.
[88] B. N. Brockhouse, "Lattice vibrations of semiconductors by neutron spectrometry", *Journal of the Physics and Chemistry of Solids*, vol. 8 (1959), pp. 400–405 and 421–422.
[89] B. N. Brockhouse and P. K. Iyengar, "Normal modes of germanium by neutron spectrometry", *Phys. Rev.*, vol. 111, no. 3 (Aug. 1, 1958), pp. 747–754. (Erratum: vol. 113, no. 6 (1959), p. 1696.)
[90] B. N. Brockhouse, "Lattice vibrations in silicon and germanium", *Phys. Rev. Lett.*, vol. 2, no. 6 (Mar. 15, 1959), pp. 256–258.
[91] B. N. Brockhouse, "Slow neutron spectroscopy and the grand atlas of the physical world", *Reviews of Modern Physics*, vol. 67, no. 4 (Oct. 1995), pp. 735–751.

[92] C. Cercignani, *Ludwig Boltzmann: The Man who Trusted Atoms*, Oxford University Press, Oxford, 1998, pp. 1–329.

[93a] E. Fermi, "Sulla quantizzazione del gas perfetto monoatomico", *Atti della Reale Accademia Nazionale dei Lincei*, vol. 3 (1926), pp. 145–149. (The title after translation from Italian to English is "Quantization of the monatomic perfect gas". The name of the journal is quite frequently known as Rendiconti Lincei. This paper can also be found in the *Enrico Fermi Collected Papers*, volume I, The University of Chicago Press, Chicago, 1962.)

[93b] E. Fermi, "Zur quantelung des idealen einatomigen gases", Zeitschrift fur Physik, vol. 36, no. 11–12 (1926), pp. 902–912. (The title after translation from German to English is "Quantization of the ideal monatomic gas". This paper can also be found in the *Enrico Fermi Collected Papers*, volume I, The University of Chicago Press, Chicago, 1962.)

[94] P. A. M. Dirac, "On the theory of quantum mechanics", *Proceedings of the Royal Society A*, vol. 112, no. 762 (Oct. 1, 1926), pp. 661–677.

[95] E. Bretscher and J. D. Cockcroft, "Enrico Fermi 1901–1954", *Biographical Memoirs of Fellows of the Royal Society*, vol. 1 (Nov. 1955), pp. 69–78.

[96] R. H. Dalitz and R. Peierls, "Paul Adrien Maurice Dirac 8 August 1902–20 October 1984", *Biographical Memoirs of Fellows of the Royal Society*, vol. 32 (Dec. 1986), pp. 139–185.

[97] A. Einstein, "Uber die von der moleckularkinetischen theorie der warme geforderte bewegung von in ruhenden flussigkeiten suspendierten teilchen", *Annalen der Physik*, vol. 17, no. 3 (18 July 1905), pp. 549–560. (Note: The title translated from German into English is "On the movement of small particles suspended in a stationary liquid demanded by the molecular-kinetic theory of heat".)

[98] A. Einstein, "Zur theorie der Brownschen bewegung", *Annalen der Physik*, vol. 19, no. 2 (8 Feb. 1906), pp. 371–381. (Note: The title translated from German into English is "Theory of the Brownian motion".)

[99] A. Einstein, *Investigations on the Theory of the Brownian Movement*, edited by R. Furth, translated into English by A. D. Cowper, Dutton, New York, 1926.

[100] J. Czochralski, "Ein neues verfahren zur messung der kristallisationsgeschwindigheit der metalle", *Zeitschrift für Physikalische Chemie* (*Z. Phys. Chemie*), vol. 92 (1918), pp. 219–221. (Note: The title translated from German into English is "A new method for the measurement of the crystallization rate of metals".)

[101] J. Evers, P. Kuflers, R. Staudigl and P. Stauhofer, "Czochralski's creative mistake: A milestone on the way to the gigabit era", *Angew. Chem. Int. Ed.*, vol. 42 (2003), pp. 5684–5698.

[102] G. K. Teal and J. B. Little, "Growth of germanium single crystals", *Phys. Rev.*, vol. 78 (1950), p. 647.

[103] J. B. Little and G. K. Teal, "Production of germanium rods having longitudinal crystal boundaries", US Patent 2,683,676, filed in 1950 and awarded in 1954.

[104] G. K Teal, "Single crystals of germanium and silicon—Basic to the transistor and integrated circuit", *IEEE Trans. Electron Dev.*, vol. 23, no. 7 (July. 1976), pp. 621–639.

[105] J. A. Burton, R. C. Prim and W. P. Slichter, "The distribution of solute in crystals grown from the melt, Part I theoretical", *J. Chem. Phys.*, vol. 21, no. 11 (Nov. 1953) pp. 1987–1991.

[106] J. A. Burton, E. D. Kolb, W. P. Slichter and J. D. Struthers, "Distribution of solute in crystals grown from the melt, Part II experimental", *J. Chem. Phys.*, vol. 21, no. 11 (Nov. 1953) pp. 1991–1996.

[107] E. E. Haller, W. L. Hansen and F. S. Goulding, "Physics of ultra-pure germanium", *Advances in Physics*, vol. 30, no. 1 (Jan./Feb. 1981), pp. 93–138.

[108] D. T. J. Hurle and P. Rudolph, "A brief history of defect formation, segregation, faceting, and twinning in melt-grown semiconductors", *Journal of Crystal Growth*, vol. 264, no. 4 (31 March 2004) pp. 550–564.

[109] *50 Years Progress in Crystal Growth: A Reprint Collection*, edited by R. S. Feigelson, Elsevier, Amsterdam, 2004, pp. 1–240.

[110] G. Bemski, "Quenched-in recombination centers in silicon", *Physical Review*, vol. 103, no. 3 (Oct. 1958) pp. 567–569.

[111] G. Bemski and J. D. Struthers, "Gold in silicon", *Journal of the Electrochemical Society*, vol. 105, no. 10 (Oct. 1958) pp. 588–591.

[112] S. J. Silverman and J. B. Singleton, "Technique for preserving lifetime in diffused silicon", *Journal of the Electrochemical Society*, vol. 105, no. 10 (Oct. 1958) pp. 591–594.

[113] A. Goetzberger and W. Shockley, "Metal precipitates in silicon p-n junctions", *J. Appl. Phys.*, vol. 31, no. 10 (Oct. 1960) pp. 1821–1824.

[114] S. W. Ing Jr., R. E. Morrison, L. L. Alt and R. W. Aldrich, "Gettering of metallic impurities from planar silicon diodes", *Journal of the Electrochemical Society*, vol. 110, no. 6 (June 1963) pp. 533–537.

[115] S. K. Ghandhi and F. L. Thiel, "The properties of nickel in silicon", *Proc. IEEE*, vol. 57, no. 9 (Sept. 1969) pp. 1484–1489.

[116] J. S. Kang and D. K. Schroder, "Gettering in silicon", *J. Appl. Phys.*, vol. 65, no. 8 (15 April 1989) pp. 2974–2985.

[117] K. Graff, *Metal Impurities in Silicon-Device Fabrication*, Springer, New York, 2000, pp. 1–268.

[118] V. A. Perevostchikov and V. D. Skoupov, *Gettering Defects in Semiconductors*, Springer, New York, 2005, pp. 1–386.

[119] B. Lojek, *History of Semiconductor Engineering*, Springer, New York, 2007, pp. 111–113.

[120] J.-C. Chou and H.-Y. Yang, "Study on the optoelectronic properties of amorphous selenium-based photoreceptors", *Optical and Quantum Electronics*, vol. 32, no. 3 (March 2000), pp. 249–261.

[121] S. A. Mahmood, M. Z. Kabir, O. Tousignant, H. Mani. J. Greenspan and P. Botka, "Dark current in multilayer amorphous selenium X-ray imaging detectors", *Appl. Phys. Lett.*, vol. 92, no. 22 (2 June 2009), pp. 223506–1 to 223506–3.

[122] L. O. Grondahl, "The copper-cuprous-oxide rectifier and photoelectric cell", *Reviews of Modern Physics*, vol. 5, no. 2 (April 1933), pp. 141–168.

[123] W. H. Brattain, "The copper oxide rectifier", *Reviews of Modern Physics*, vol. 23, no. 3 (July 1951), pp. 203–212.

[124] J. Mizuguchi, "Observation of the potential barrier in cuprous oxide rectifier with scanning electron microscopy using beam induced current", *Jpn. J. Appl. Phys.*, vol. 15, no. 5 (May 1976), pp. 907–908.

[125] L. Wang and M. Tao, "Fabrication and characterization of p-n homojunctions in cuprous oxide by electrochemical deposition", *Electrochemical and Solid-State Letter*, vol. 10, no. 9 (Sep. 2007), pp. H248–H250.

[126] K. Han and M. Tao, "Electrochemically deposited p-n homojunction cuprous oxide solar cells", *Solar Energy Materials and Solar Cells*, vol. 93, no. 1 (Jan. 2009), pp. 153–157.

[127] V. A. Johnson, *Karl Lark-Horovitz: Pioneer in Solid-State Physics*, Pergamon Press Ltd., Oxford, UK, 1969, pp. 1–289.

[128] P. W. Henriksen, "Solid state physics research at Purdue", *Osiris*, second series, vol. 3 (1987), pp. 237–260.

[129] C. Claeys and E. Simeon, *Germanium-based Technology from Materials to Devices*, Elsevier, Oxford, UK, 2007, pp. 1–449.

[130] F. Seitz and N. G. Einspruch, *Electronic Genie: The Tangled History of Silicon*, University of Illinois Press, Urbana and Chicago, USA, 1998.

[131] L. Hoddeson, E. Braun, J. Teichmann and S. Weart, *Out of the Crystal Maze: Chapters from the History of Solid-State Physics*, Oxford University Press, New York, USA, 1992.

[132] B. Lojek, *History of Semiconductor Engineering*, Springer, New York, 2007.

[133] K. Murphy, "Happy birthday, Fairchild", *IEEE Spectrum*, vol. 44, no. 12 (December 2007), p. 8.

[134] L. Berlin, *The Man Behind the Microchip: Robert Noyce and the Invention of Silicon Valley*, Oxford University Press, Oxford, 2005, pp. 1–402.

[135] T. R. Reid, *The Chip: How Two Americans Invented the Microchip and Launched a Revolution*, Random House, New York, 2001, pp. 1–309.

[136] C. Lecuyer, *Making Silicon Valley: Innovation and the Growth of High Tech, 1930–1970*, MIT Press, Cambridge, Masschusetts, USA, 2006, pp. 1–393.

[137] N. F. Mott, "Note on the contact between a metal and an insulator or semiconductor", *Proceedings of the Cambridge Philosophical Society*, vol. 34 (1938), pp. 568–572.

[138] F. A. Kroger, G. Diemer and H. A. Klasens, "Nature of an Ohmic metal-semiconductor contact", *Physical Review*, vol. 103, no. 2 (15 July 1956), p. 279.

[139] A. Y. C. Yu, "The metal-semiconductor contact: an old device with a new future", *IEEE Spectrum*, vol. 7, no. 3 (March 1970), pp. 83–89.

[140] A. Y. C. Yu, "Electron tunneling and contact resistance of metal-silicon contact barriers", *Solid-State Electron.*, vol. 13, no. 2 (Feb. 1970), pp. 239–247.

[141] A. Dimoulas, A. Toriumi and S. E. Mohney, " Source and drain contacts for germanium for III–V FETs for digital logic", *MRS Bulletin*, vol. 34, no. 7 (July 2009), pp. 522–529.
[142] E. Merzbacher, "The early history of quantum tunneling", *Physics Today*, vol. 55, no. 8 (Aug. 2002) pp. 44–49.
[143] R. H. Fowler and L. W. Nordheim, "Electron emission in intense electric fields", *Proceedings of the Royal Society of London A*, vol. 119, no. 781 (May 1, 1928) pp. 173–181.
[144] E. O. Johnson, "The insulated-gate field-effect transistor — a bipolar transistor in disguise", *RCA Review*, vol. 34, no. 1 (March 1973), pp. 80–94.
[145] R. E. Thomas, R. H. Johnston and A. R. Boothroyd, "Negative resistance in a transistor under punch-through conditions", *Proc. IEEE*, vol. 54, no. 1 (Jan. 1966), pp. 84–85.
[146] P. J. Ward and K. D. Perkins, "Current/voltage relations in punchthrough transistors", *Electron. Lett.*, vol. 10, no. 18 (5th Sept. 1974), pp. 374–375.
[147] D. Frohman-Bentchkowsky and A. S. grove, "Conductance of MOS transistors in saturation", *IEEE Trans. Electron Dev.*, vol. 16, no. 1 (Jan. 1969), pp. 108–113.
[148] R. A. Stuart and W. Eccleston, "Punchthrough currents in short-channel M.O.S.T. devices", *Electron. Lett.*, vol. 9, no. 25 (13th Dec. 1973), pp. 586–588.
[149] H. Christensen, "Surface conduction channel phenomena in germanium" *Proc. IRE*, vol. 42, no. 9 (Sep. 1954), pp. 1371–1376.
[150] R. H. Kingston, "Review of germanium surface phenomena", *J. Appl. Phys.*, vol. 27, no. 2 (Feb. 1956) pp. 101–114. (MIT)
[151] J. H. Forster and H. S. Veloric, "Effect of variations in surface potential on junction characteristics", *J. Appl. Phys.*, vol. 30, no. 6 (June 1959) pp. 906–914.
[152] C.-T. Sah, "Effect of surface recombination and channel on P-N junction and transistor characteristics", *IRE Trans. Electron Dev.*, vol. 9, no. 1 (Jan. 1962), pp. 94–108.
[153] A. S. Grove and D. J. Fitzgerald, "Surface effects on silicon P-N junctions: The origin of anomalous channel currents", *IEEE Trans. Electron Dev.*, vol. 12, no. 9 (Sept. 1965), pp. 508–509.
[154] A. S. Grove and D. J. Fitzgerald, "Surface effects on p-n junctions: characteristics of surface space-charge regions under non-equilibrium conditions", *Solid-State Electron.*, vol. 9, no. 8 (Aug. 1966) pp. 783–806.
[155] A. S. Grove and D. J. Fitzgerald, "The origin of channel currents associated with p^+ regions in silicon", *IEEE Trans. Electron Dev.*, vol. 12, no. 12 (Dec. 1965), pp. 619–626.
[156] A. S. Grove, O. Leistiko Jr. and W. W. Hooper, "Effect of surface fields on the breakdown voltage of planar silicon p-n junctions", *IEEE Trans. Electron Dev.*, vol. 14, no. 3 (Mar. 1967), pp. 157–162.
[157] A. S. Grove, *Physics and Technology of Semiconductor Devices*, Wiley, New York, 1967, Chapter 10, "Surface effects on p-n junctions", pp. 289–316.

Chapter Two

History of MOS Technology

2.1 History of the Invention of the MOS Transistor

As discussed in Chapter One, Lee de Forest (1873–1961) was an American scientist who invented vacuum tube triodes and he got US Patent 879,532 in 1908 for this invention. It is not so difficult to imagine that a similar but more compact solid state device can be fabricated. Julius Edgar Lilienfeld (1882–1963) was born in 1882 in Lemberg, Austria-Hungary (now known as Lviv, Ukraine). He got his PhD from the Frederick William University in Berlin (now known as Humboldt University of Berlin) in 1905. He spent some years in Germany. Subsequently he moved to USA in the early 1920's. He proposed a device similar to the modern MESFET (metal semiconductor field effect transistor) in 1926.[1] A few years later, he proposed a device similar to the modern MOS transistor in 1928.[2] Lilienfeld has also been famous for inventing the electrolytic capacitor. A German scientist, Oskar Heil (1908–1994) proposed something similar. He described the possibility of controlling the resistance in a semiconducting material with an electric field in British patent 439,457 with the title "Improvements in or relating to electrical amplifiers and other control arrangements and devices" in 1935.[3] Like Lilienfeld, Heil later also ended up in USA. Especially in earlier days, people could get patents on electronic devices without actually fabricating a working device. When people actually tried to make a practical device, it was found that the reality was not simple. The actual effect experimentally observed was quite small. In the book "Crystal Fire" by Riordan and Hoddeson, there was a lengthy description regarding

Shockley's attempt to make a field effect transistor. An American scientist William Bradford Shockley (1910–1989) was born in London by American parents. He grew up in California and was educated at CalTech and MIT. He joined Bell Telephone Laboratories in 1936. In 1939, Shockley attempted to build a solid-state amplifier using cuprous oxide (Cu_2O). At that time, vacuum tubes dominated electronics; however, vacuum tubes take up a lot of space, consume a lot of power and are mechanically fragile. As discussed in Chapter One, cuprous oxide was one of the earliest known semiconductors and was once popularly used for rectifiers. He did not manage to observe any significant effect out of his attempt to build a solid-state field effect transistor based on cuprous oxide. World War II interrupted his work but it was resumed in 1945. According to the book "Crystal Fire" by Riordan and Hoddeson, in 1945, Shockley made another attempt.[4] "With the help of co-workers, he obtained a small ceramic cylinder onto which a thin film of silicon had been deposited. Attaching a 90-volt battery to its two ends, they could detect only the tiniest current trickling through the silicon film. Then they applied a thousand volts across a narrow gap — less than a millimeter wide — above the silicon layer. With such a powerful electric field pulling electrons to the surface, a surge of current should have occurred, but none was observed." Shockley made more attempts without success. "Thoroughly mystified by his failures, he abandoned the effort and turned to other fancies."

As discussed in Chapter One, Bardeen and Brattain invented the point contact transistor in 1947 while Shockley invented the junction transistor somewhat later. Both the point contact transistor and the junction transistor belong to the family of bipolar transistors. The theory of bipolar transistors is actually much more complicated and difficult to understand than that of field effect transistors. However, it will take more years for a workable field effect transistor to be demonstrated.

Igor Yevgenyevich Tamm (1895–1971) was a famous Soviet scientist. He is also known as Igor Evgenevich Tamm and so abbreviated as I. E. Tamm. One of his scientific achievements was to propose a concept known as surface states in 1932.[5] He is also famous because he got Nobel Prize in physics in 1958 for the interpretation of the Cherenkov-Vavilov effect. Surface states according to Tamm's theory is known as Tamm states. Later in 1939, Shockley proposed his theory on surface states.[6] Surface states according to Shockley's theory is known as Shockley states. Historically surface states that arise as solutions to the Schrödinger equation in the framework of the nearly-free electron approximation for clean and ideal surfaces, are called Shockley states. Shockley states are thus states that arise due to the change in the electron potential associated solely with the crystal termination. This approach is suited to describe normal metals and some narrow gap semiconductors. Surface states that are calculated in the framework of a tight-binding model are often called Tamm states. In the tight binding approach the electronic wave functions are usually expressed as Linear Combinations of Atomic Orbitals (LCAO). In contrast to the nearly-free electron model used to describe the Shockley states, the Tamm states are suitable to describe also transition metals and wide gap semiconductors. Nowadays, we know that Shockley's failed attempts in 1939 and 1945 were caused by surface states.

After the invention of practical bipolar transistors, field effect transistors eventually debuted in 1953, but in the form of a junction field effect transistor (JFET) instead of MOSFET. In 1952, Shockley published a theoretical paper on JFET.[7] This was followed by a paper from Dacey and Ross with experimental data in 1953.[8] Ian Munro Ross (1927–2013) was an early pioneer in transistors; in addition, he has served for 12 years as President of Bell Laboratories. At this point, the readers should note that there is a related device known as the MESFET (metal semiconductor field effect transistor) invented by Carver Andress Mead (1934–).[9] Commercial Si JFETs are readily available but there is no commercial Si MESFET readily available in the market. However, GaAs MESFETs have been available commercially for many years.

Martin M. (John) Atalla (1924–2009) was born in Port Said, Egypt. After he got his PhD degree from Purdue University, he joined Bell Laboratories in 1949. One of his inventions is the first practical MOS transistor based on Si. Atalla investigated the application of a surface silicon dioxide film to passivate the surface states at the silicon surface.[10] Atalla then suggested that a field effect transistor be built of metal-oxide-silicon. He assigned the task to Dawon Kahng, a scientist in his group. Dawon Kahng (1931–1992) was born in Seoul, Korea when Korea was within the Japanese Empire. In 1955, he moved to USA. Subsequently, he joined Bell Telephone Laboratories. Eventually, Attalla and Kahng announced their successful MOSFET at a 1960 conference.[11] Kahng published a technical memorandum of Bell Laboratories issued on January 16, 1961 with the title of "Silicon-silicon dioxide surface device".[12] Kahng obtained a patent for MOSFET in 1963.[13] Many years later in 1976, Kahng published an article in IEEE Transactions on Electron Devices for his historical invention.[14] Besides the use of the SiO_2/Si system, hydrogen annealing was another important technical development to lower the density of surface states. It was first reported by P. Balk in a 1965 conference[15a]; Balk subsequently followed up on his pioneering work, for example, in Ref. 15b. More details regarding this annealing process can be found in the book "MOS Physics and Technology" by Nicollian and Brews.[16] In 1998, Arns reviewed the early history of the MOSFET device.[17]

As discussed above, surface states have been a serious problem in early MOS devices. In addition, there was another serious problem: the stability of the electrical characteristics of early MOS devices. One serious problem was that the threshold voltage was unstable because of the drifting of alkaline ions, for example, Na^+, K^+, Li^+, etc. Usually sodium ions may be more important than other alkaline ions. In a short article published in the May 1991 issue of IEEE Spectrum, Riezenman pointed out that Frank Wanlass (1933–2010) had played a crucial role to solve this serious problem[18]; he also pointed out that Wanlass invented CMOS. Frank Marion Wanlass was born in 1933 and passed away in 2010. After he got his PhD from the University of Utah, he joined Fairchild in 1962. Frank Wanlass used electron beam evaporation of the metal gate to get stable MOS transistors instead of heating the metal with a refractory metal filament which may have been contaminated by sodium. However, another article by Basset pointed out that it was Snow who solved this problem.[19] The book by Lojek[20] pointed out that Wanlass joined Fairchild Semiconductor in 1962 and then quitted in 1964; Snow continued to follow up on some work by Wanlass. Therefore, it may be Wanlass who started the breakthrough but Snow performed the follow up work. Finally, it was Snow and his colleagues who published a classical paper on this subject in the *Journal of Applied Physics*.[21] Snow and his colleagues did not mention the name of Wanlass in their paper.

2.2 Control of the Threshold Voltage and the Threshold Adjust Implant

MOS transistors can be based on the movement of electrons in n-channel MOS transistors; conversely, MOS transistors can be based on the movement of holes in p-channel MOS transistors. It can be easily imagined that n-channel devices are faster than p-channel devices because the electron mobility of silicon is higher than the hole mobility of silicon. In addition, MOS transistors can be made as depletion-mode or enhancement-mode transistors. For digital circuits, enhancement-mode MOS transistors are usually preferred as switching transistors. In the early days of MOS technology, it was easier to make p-channel enhancement-mode MOS transistors with a reasonable threshold voltage than n-channel enhancement-mode MOS transistors; at that time, n-channel enhancement-mode MOS transistors tended to have insufficient threshold voltage. Because of this, PMOS digital integrated circuit technology based on p-channel enhancement-mode transistors was once a popular choice. For example, the first microprocessor Intel 4004 was fabricated by PMOS technology. A PMOS inverter can be made by a p-channel enhancement-mode transistor with a load resistor. Frequently the load resistor was replaced by another p-channel enhancement-mode transistor with the drain and gate shorted together to the negative power supply voltage. For PMOS technology, the starting substrate was usually a lightly-doped n-type (100) Si wafer. The major steps involved in the fabrication of PMOS transistors can be found in the article by Schmid.[22] Nowadays, PMOS process is more or less obsolete for ordinary applications. The only known advantage of p-channel devices over n-channel devices is for applications in an environment with high energy radiation. For example, p-channel CCD can be superior to n-channel CCD for special applications involving high energy radiation.[23] NMOS process usually starts from p-type Si wafers. As we shall see later, CMOS process can start from n-type or p-type Si wafers.

As discussed above, the first PMOS integrated circuit technology used p-channel enhancement-mode transistor as the switching transistor and another p-channel enhancement-mode transistor as the load resistor. Later, the p-channel enhancement-mode transistor serving as the load was replaced by a p-channel depletion-mode transistor, as discussed by Borel *et al.*[24] For the enhancement load, the drain and gate are connected together; however, for the depletion load, the gate and source are connected together, as discussed by Borel *et al.*[24] The advantage of this approach can be explained as follows. The switching the PMOS inverter is similar to charging and discharging a parasitic capacitor connected from the inverter output to the ground. When the inverter input is at "logic 0", the parasitic capacitor will be charged from "logic 0" to "logic 1". For an enhancement load, the charging process will become slow when the output voltage is close to the power supply voltage. For a depletion load, the charging process will be faster when the output voltage is close to the power supply voltage when compared to the enhancement load. The same theory can be applied to NMOS inverter.

As discussed above, in the early days of MOS technology, it was easier to make p-channel enhancement mode transistor compared to n-channel enhancement mode transistor. This was because the threshold voltage tended to be low or even negative for early n-channel enhancement mode MOS transistor. Subsequently, this problem was solved by using a boron threshold adjust implant. Once this problem was solved, NMOS digital integrated circuit technology quickly dominated over the older PMOS technology because the electron mobility is larger than the hole mobility and so the n-channel device is naturally superior to the p-channel device. According to Lojek's book,[25] Kenneth G. Aubuchon of Hughes Research Laboratories pioneered the use of ion implantation to set the threshold voltage of MOS transistors.[26] He presented his results in a conference in Grenoble, France in 1969. This was followed up by MacPherson.[27,28] In his papers, MacPherson quoted Aubuchon's 1969 conference paper. MacPherson used a boron implant to reduce the magnitude of threshold voltage of p-channel transistors.[27] More theoretical discussion regarding the shifting of threshold voltage of p-channel MOS transistor was given by Sigmon and Swanson in 1973.[29] After the successful application of ion implantation to the control of the threshold voltage of p-channel transistors, boron implant was applied to n-channel transistors to increase the magnitude of the threshold voltage; thus the problem of low threshold voltage of n-channel enhancement-mode transistors can be solved by a boron threshold adjust implant.[30] More discussion on the history of the threshold adjust implant was found in the book by Lojek. After the problem of low threshold voltage of Si n-channel enhancement-mode transistors was solved, NMOS digital integrated circuit technology quickly dominated over the PMOS digital integrated circuit technology such that nowadays PMOS technology basically disappears. As discussed above, PMOS technology with depletion load is superior to PMOS technology with enhancement load; similarly, NMOS technology with depletion load is superior to NMOS technology with enhancement load. Eventually, NMOS technology using enhancement-mode n-channel transistor as the switching transistor with depletion-mode n-channel transistor serving like a load resistor became the standard NMOS technology.

2.3 Invention of CMOS

When the number of transistors increases, the heat generated in NMOS digital integrated circuits can be a serious problem. This problem can be solved by CMOS digital integrated circuit technology. CMOS stands for "Complementary metal-oxide-semiconductor". It was invented by Frank Wanlass (1933–2010), who was also a pioneer working on MOS threshold voltage instability problem as discussed above.[31,32] Wanlass got US Patent 3,356,858 for his CMOS invention; since Wanlass worked for Fairchild at the time of his invention, the patent was owned by Fairchild. The first CMOS integrated circuits were made by RCA in 1968 by a group led by Albert Medwin (1925–).

Originally a low-power but slow alternative to TTL (transistor transistor logic), CMOS found early adopters in the watch industry and in other fields where battery life was more important than speed. Some 25 years later, CMOS has become the predominant technology in digital integrated circuits. This is essentially because area occupation, operating speed, energy efficiency and manufacturing costs have benefited and continue to benefit from the geometric downsizing that comes with every new generation of semiconductor manufacturing processes. In addition, the simplicity and comparatively low power dissipation of CMOS circuits have allowed for integration densities not possible on the basis of bipolar junction transistors. Standard discrete CMOS logic functions were originally available only in the 4000 series (RCA "COS/MOS") integrated circuits. Later many functions in the 7400 series began to be fabricated by CMOS, NMOS or BiCMOS technologies.

Early CMOS circuits were very susceptible to damage from electrostatic discharge (ESD). Subsequent generations were thus equipped with sophisticated protection circuitry that helps absorb electric charges with no damage to the fragile gate oxides and PN-junctions. Still, antistatic handling precautions for semiconductor devices continue to be followed to prevent excessive energies from building up. Manufacturers recommend using antistatic precautions when adding a memory module to a computer, for instance.

On the other hand, early generations such as the 4000 series that used aluminum as a gate material were extremely tolerant of supply voltage variations and operated anywhere from 3 to 18 volts DC. For many years, CMOS logic was designed to operate from the industry-standard of 5 V imposed by TTL. By 1990, lower power dissipation was usually more important than easy interfacing to TTL, and CMOS voltage supplies began to drop along with the geometric dimensions of the transistors. Lower voltage supplies not only saved power, but allowed thinner, higher performance gate insulators to be used. Some modern CMOS circuits operate from voltages below 1 V.

In the early fabrication processes, the gate electrode was made of aluminum. Later CMOS processes switched to polycrystalline silicon ("polysilicon"), which can better tolerate the high temperatures used to anneal the silicon after ion implantation. (Note: The temperature needed is about 1000°C.) This means that the gate can be put on early in the process and then used directly as an implant mask producing a self aligned gate (gates that are not self aligned require overlap which increases device size and stray capacitance). Considerable research that has gone into using metal gates has led to the announcement of their use in conjunction with the replacement the silicon dioxide gate dielectric with a high-k dielectric material to combat increasing leakage currents.

CMOS can be used for both digital integrated circuits and analog integrated circuits. For analog CMOS, it can be used for low frequency applications, for example, CMOS operational amplifiers. The first CMOS operational amplifier was CA3130 from RCA. CA3130 uses p-channel MOS transistors for the input and has a CMOS output stage. (Note: After RCA stopped functioning, CA3130 was still supplied by Harris. Afterwards, CA3130 is currently still supplied by Intersil.) For analog CMOS, it can also be used for high frequency applications, for example, CMOS microwave amplifiers. Digital CMOS and analog CMOS can be included in the same integrated circuit; this is known as "mixed signal" technology.

2.4 Invention of the Silicon Gate Process

2.4.1 *Migration from metal gate to silicon gate*

As discussed above, the invention of the silicon gate process was an important milestone of MOS fabrication process. Before this, a metal gate process was used. A mask was involved in defining the source and drain diffusion. Another mask was involving in defining the gate metal. The gate metal was usually aluminum. The gate metal process was after the source/drain formation process. To allow for misalignment between these two masking steps, a large overlap between the metal and the drain (or the source) was necessary, resulting in a large overlap capacitance, which can degrade switching speed.

At this point, the author would like to introduce a key concept known as the Miller effect, which is named after John Milton Miller (1882–1962). Miller was born in Hanover, Pennsylvania, USA. He received his PhD from the Yale University in 1915. In June 1919, Miller wrote an important paper and got it published by the National Bureau of Standards, USA.[33] Nowadays, this seminal paper can be downloaded easily via the Internet. At that time, transistors were not invented yet and the dominant electronic device was the vacuum tube triode; as discussed in Chapter One, the vacuum tube triode was patented by Lee de Forest in 1908. According to the Miller effect, the grid to anode capacitance has the greater negative impact on the speed compared to the grid to cathode capacitance. Hence, it is more important to reduce the grid to anode capacitance. As discussed in Chapter One, the vacuum tube pentode was patented by Holst and Tellegen in 1934. Pentodes tend to have smaller grid to anode capacitance than triodes and so are better devices. The same idea can be extended from vacuum tubes to bipolar transistors and MOS transistors. The idea of Miller has been further developed and incorporated into a modern text book like that written by Millman and Halkias.[34] For MOS transistors, both the gate to source overlap capacitance and the gate to drain overlap capacitance can degrade the speed. However, according to the Miller effect, the gate to drain overlap capacitance has the greater negative impact on the speed compared to the gate to source overlap capacitance. In general, MOS transistors are symmetrical such that the gate to drain overlap capacitance and the gate to source overlap capacitance are equal. To increase the speed of MOS transistors, the overlap capacitance between the gate and the drain/source has to be reduced.

Bower *et al.* (Hughes Aircraft Company) developed a self-aligned process using the gate metal as a mask for the self-aligned implant of the drain and source.[35] They used 4000 Å aluminum gate as a mask for boron implant into their p-channel MOS transistor. However, the problem of using aluminum as an ion implant mask is that the melting point is about 660°C, which is relatively low, such that a high temperature anneal cannot be performed. Sarace *et al.* (Bell Laboratories) developed a self-aligned process using polycrystalline silicon as the gate metal.[36,37] In this process, amorphous silicon deposited by evaporation was turned into polycrystalline silicon by a high temperature annealing process. This technology was quickly adopted by other companies, for example, Fairchild and Intel. Vadasz *et al.* (Intel) published an article in IEEE Spectrum in 1969 about this new technology.[38] Faggin and Klein (Fairchild) made a comparison of evaporated silicon, sputtered silicon and CVD silicon.[39] (Note: CVD is chemical vapor deposition.) They found that CVD silicon seems to be the best option. The readers should note that silicon deposited by evaporation or sputtering may have step coverage problem; a silicon film deposited by CVD has no step coverage problem. Indeed, the first silicon IC using the silicon gate technology was the Fairchild 3708, an 8-bit analog multiplexer with decoding logic. Some more discussion was given by Lee and Mayer in their review article "Ion-implanted semiconductor devices" in 1974.[40] Subsequently, Federico Faggin (1941–) quitted Fairchild to join Intel. Silicon gate technology was indeed adopted by Intel for the first microprocessor IC, Intel 4004.[41,42] As discussed above, the self-aligned silicon gate technology can provide higher speed compared with the non-self-aligned aluminum gate technology.

Besides Federico Faggin, the chief designers of the Intel 4004 chip were Ted Hoff (1937–) of Intel and Masatoshi Shima (1943–) of Busicom. In terms of process technology, Intel 4004[41,42] was based on 10 μm PMOS technology using enhancement load; as discussed above, the silicon gate technology was also adopted. Subsequently, PMOS technology was replaced by NMOS technology. As discussed above, NMOS technology is inherently faster than PMOS technology. In addition, the enhancement load was replaced by depletion load. As discussed above, the depletion load is superior to the enhancement load. For some years, the standard MOS digital integrated circuit technology was using n-channel enhancement-mode MOS transistors as switching transistors and n-channel depletion-mode transistors as loading resistors.

As discussed above, the first microprocessor Intel 4004 was fabricated by PMOS technology with the silicon gate technology adopted. Thus the silicon gate in Intel 4004 was actually p^+-poly. As discussed above, MOS integrated circuit technology migrated from PMOS to NMOS with the silicon gate technology still adopted. Thus the silicon gate in NMOS integrated circuits was actually n^+-poly. As the level of integration became more complicated, MOS integrated circuit technology migrated from NMOS to CMOS with the silicon gate technology still adopted. Then there was a question regarding the conductivity type of the silicon gate material adopted. In the very early days of CMOS technology, metal gate (usually Al gate) was used, as discussed by Schmid.[22] The major process steps involved in old Al gate CMOS technology has been discussed by Schmid.[22] Subsequently, Al gate CMOS technology migrated to CMOS technology based on the self-aligned silicon gate approach. In the early days of CMOS technology using the silicon gate approach, the silicon gate was n^+-poly for both the n-channel MOS transistor and the p-channel MOS transistor. The n-channel MOS transistor with n^+-poly gate was a surface channel MOS transistor while the p-channel MOS transistor with n^+-poly gate was a buried channel MOS transistor. This is because the magnitude of the threshold voltage of a p-channel MOS transistor with n^+-poly gate tends to be too high. A p-type counter doping at the gate oxide/channel interface was quite frequently used as a threshold adjust implant in addition to the n-type doping due to the n-well, resulting in a buried channel MOS transistor.[43] In this way, the gate becomes farther away from the channel, resulting in degraded on current and also subthreshold swing. (Note: Degraded subthreshold swing implied higher off current for the same threshold voltage.) Subsequently, CMOS technology with n^+-poly for both the n-channel MOS transistor and the p-channel MOS transistor migrated to CMOS technology with n^+-poly for the n-channel MOS transistor and p^+-poly for the p-channel MOS transistor. This was known as the "dual gate" CMOS technology.[44,45] The n-channel MOS transistor with n^+-poly gate was a surface channel MOS transistor while the p-channel MOS transistor with p^+-poly gate was also a surface channel MOS transistor. For a p-channel MOS transistor with p^+-poly gate, the magnitude of the threshold voltage tends to be too low. An n-type counter doping at the gate oxide/channel interface was quite frequently used as a threshold adjust implant in addition to the n-type doping due to the n-well, resulting in a surface channel MOS transistor.[46] In general, a p-channel MOS transistor with p^+-poly gate tends to perform better than a p-channel MOS transistor with n^+-poly gate.[47] However, a p-channel MOS transistor with p^+-poly gate may suffer from the problem of "boron penetration".[48] "Boron penetration" is the problem of boron from the p^+-poly gate goes through the gate oxide into the channel in a p-channel MOS transistor with p^+-poly gate. This problem is usually solved by using a nitride gate oxide. In general, silicon nitride is a much better diffusion barrier than silicon dioxide; adding nitrogen to silicon dioxide usually will make a better diffusion barrier. This can be done, for example, by a nitrous oxide (N_2O) anneal.[48] Similarly, a nitric oxide (NO) anneal can also be used.[49]

After this, CMOS technology migrated from using n^+-poly gate for the n-channel MOS transistor and p^+-poly gate for the p-channel MOS transistor to using n^+-poly-SiGe gate for the n-channel MOS transistor and p^+-poly-SiGe gate for the p-channel MOS transistor. This is because of a new issue known as "polysilicon depletion" or "poly depletion". The physical origin of this problem is that the free electron concentration of n^+-poly gate or the free hole concentration of p^+-poly gate is limited. Further increase of doping concentration in the poly gate may run into solid solubility limit of the dopant atoms in silicon such that the dopant activation can be poor in very highly doped poly. Another effect is the quantum mechanical effect happening in the inversion region at the interface of the gate dielectric and silicon for enhancement-mode MOS transistors. For n-channel MOS transistors, the electrons appear to be slightly below the interface of the gate dielectric and silicon; similarly, for p-channel MOS transistors, the holes appear to be slightly below the interface of the gate dielectric and silicon. The net effect of polysilicon depletion and the quantum mechanical effect is that the gate oxide seems to be "thicker". For example, the physical gate oxide thickness can be 1.6 nm; however, electrically, the gate oxide thickness looks like 2.6 nm. Some relevant references are given in this book as references.[50–54] The advantage of poly-SiGe gate is that the dopant activation is better for heavily boron doped p^+-poly-SiGe compared to heavily boron doped p^+-poly-Si. This was pointed out, for example, by King *et al.* in 1994.[55] As shown in Fig. 2.1(a), the boron dopant activation can be improved from about 40 % to about 80 % by switching from poly-Si to poly-SiGe. In addition, as shown in Fig. 2.1(b), the hole mobility is also improved by switching from poly-Si to poly-SiGe. Subsequently, poly-SiGe based CMOS has been adopted, for example, by Ponomarev *et al.*,[56] etc.

At this point, the author would like to point out that the dual gate CMOS technology may be good for logic ICs. However, dual gate CMOS technology is more complicated and so can be more costly. In DRAM technology, n^+-poly may still be used for both n-channel and p-channel devices because the technology is simpler and there is no boron penetration problem! For example, Inaba *et al.* gave a report on single work function gate DRAM technology in 2002.[57] Usually, single work-function gate means using n^+-poly gate for both n-channel and p-channel transistors. However, Sim and Kim[58] discussed about using p^+-poly gate in DRAM technology in 1998. Of course, dual gate CMOS technology can also be adopted for DRAM technology but the process is more complicated and so more costly. For example, Hiura *et al.* reported on this approach in IEDM 1998.[59] It is just a matter of cost effectiveness.

2.4.2 *Back to metal gate*

The size of the MOS transistor has been scaled down from the 1960's all the way down to the 21st century. When the gate length is scaled down, the thickness of gate oxide has to be scaled down also. Eventually, the physical thickness of the gate oxide is reduced to below 2 nm. The gate oxide leakage current can become a serious concern. The dielectric constant of silicon dioxide is only 3.9. The solution of this problem is to use a gate dielectric with a dielectric constant significantly higher than that of silicon dioxide; this is known as high-k dielectric.[60–62] As mentioned earlier, poly depletion can be important especially when the gate oxide is very thin. The solution of this problem is to use a metal gate. The combination of high-k dielectric together with metal gate can bring very significant improvement to

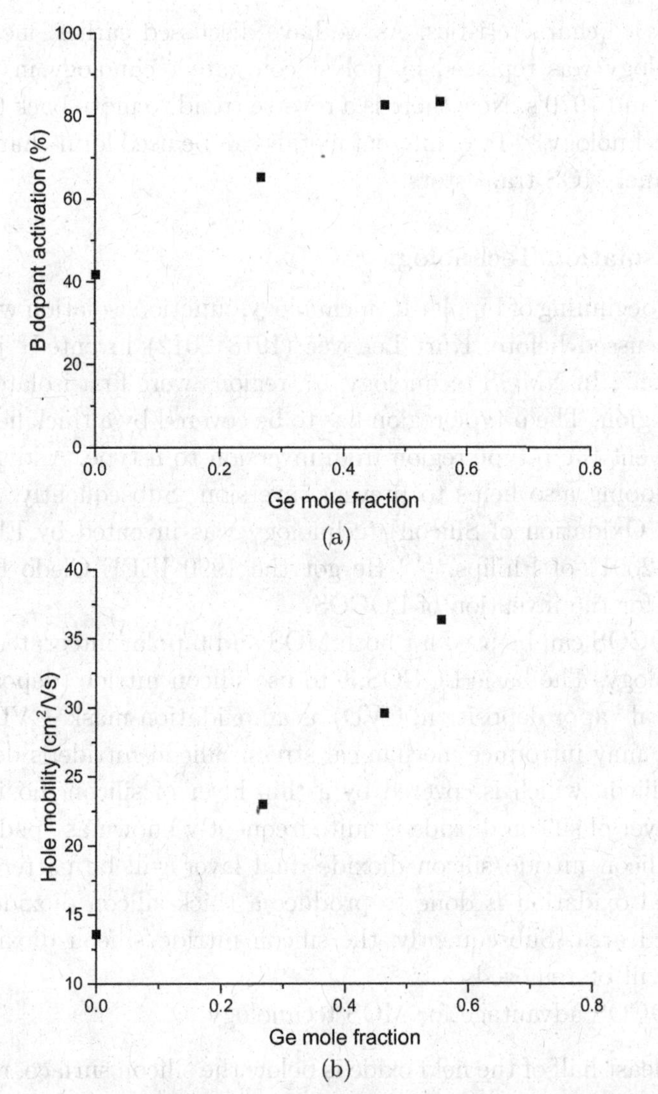

Fig. 2.1 (a) The advantage of poly-SiGe over poly-Si gate is that the % boron activation in p-type poly-SiGe is better than p-type poly-Si according to King *et al.*[25] (b) In addition, the hole mobility of p-type poly-SiGe is also better than that of p-type poly-Si according to King *et al.*[25] (Modified into 2 sub-figures from Fig. 2 in T.-J. King, J. P. McVittie, K. C. Saraswat, and J. R. Pfiester, "Electrical properties of heavily doped polycrystalline silicon-germanium films," *IEEE Trans. Electron Dev.*, vol. 41, no. 2 (Feb. 1994), pp. 228–232.)

the device characteristics. As we have discussed earlier, metal gate technology was replaced by polysilicon gate technology in the late 1960's and 1970's. Now there is a reverse trend to move back to metal gate technology.[63] Two different metals can be used for n-channel and p-channel MOS transistors.

2.5 Isolation Technology

In the beginning of bipolar IC technology, junction isolation was used. As discussed before, Kurt Lehovec (1918–2012) invented "junction isolation". In NMOS technology, n^+ regions were first isolated by p-type region. The p-type region has to be covered by a thick field thick to prevent the p-type region from inversion to n-type. Additional p-type doping also helps to prevent inversion. Subsequently, LOCOS (Local Oxidation of Silicon) technology was invented by Else Kooi (1932–2001) of Philips.[64,65] He got the 1990 IEEE Cledo Brunetti award for the invention of LOCOS.

LOCOS can be used for both MOS and bipolar integrated circuit technology. The basic LOCOS is to use silicon nitride (deposited by chemical vapor deposition, CVD) as an oxidation mask. CVD silicon nitride may introduce mechanical stress. Silicon nitride is deposited onto silicon which is covered by a thin layer of silicon dioxide; this thin layer of silicon dioxide is quite frequently known as "pad oxide". The silicon nitride/silicon dioxide dual layer will be pattered. The thermal oxidation is done to produce a thick silicon dioxide in the exposed area. Subsequently, the silicon nitride/silicon dioxide dual layer will be removed.

LOCOS advantage for MOS technology:

1. At least half of the field oxide is below the silicon surface, resulting in smoother topology. (Note: Before the advent of CMP technology, non-flat topology was always a problem!)
2. Replacing the p-region by silicon dioxide decreases parasitic capacitance, resulting in better speed.

LOCOS advantage for bipolar technology:

1. Replacing the p-region by silicon dioxide can lead to better "design rule". Now the field region can be touching the p-type base of

npn transistors used in bipolar technology. Using a p-region for isolation, the p-type field region and the p-type base have to be separated by a n-region to avoid shorting. Better "design rule" implies higher "packing density".

2. Replacing the p-region by silicon dioxide decreases parasitic capacitance, resulting in better speed.

The disadvantage of the LOCOS process is that there is always some oxidation under the silicon nitride/silicon dioxide dual layer adjacent to the exposed area. This is known as "bird's beak". When CMOS technology reached the 0.25 μm node, LOCOS was replaced by a new process known as shallow trench isolation (STI). For STI, shallow trenches are etched in silicon; these shallow trenches will be filled with silicon dioxide by CVD or some sort of modified CVD technique. For example, some sort of high density plasma chemical vapor deposition (HDPCVD) has been developed for STI applications.[66,67] Subsequently, chemical mechanical polishing (CMP) can be used for "planarization" of the non-flat surface.

For both LOCOS and STI isolation technology, an isolation implant is quite frequently needed for the purpose of the isolation of MOS transistors from each other. For example, a boron implant can be done through STI such that the p-well below STI is more strongly p-type. The design of this isolation implant can be considered part of well engineering.

When LOCOS isolation is used, MOS transistors suffer from "narrow width effect" (NWE); the magnitude of the threshold voltage becomes larger when the gate width of the MOS transistor becomes smaller. When STI isolation is used, MOS transistors suffer from "reverse narrow width effect" (RNWE); the magnitude of the threshold voltage becomes smaller when the gate width of the MOS transistor becomes smaller. For MOS transistors with very small gate lengths, an anomalous narrow width effect may be observed as follows.

N-channel surface-channel MOS transistors using shallow trench isolation (STI) is known to show reverse narrow width effect (RNWE) such that the threshold voltage becomes smaller when the channel width decreases. Lau *et al.*[68] found that by using a phosphorus deep

S/D implant in addition to an arsenic deep S/D implant, the threshold voltage first becomes larger when the channel width decreases and then later becomes smaller when the channel width further decreases for n-channel MOS transistors with very small gate lengths. Such an anomalous narrow width effect was attributed to an enhancement of TED (transient enhanced diffusion) due to Si interstitials generated by the phosphorus implant. P-channel MOS transistors show up a much stronger anomalous narrow width effect compared to n-channel MOS transistors. Such an anomalous narrow width effect was attributed to an enhancement of phosphorus and arsenic TED due to Si interstitials generated by the deep boron S/D implant.[68,69]

2.6 Drain/Source Engineering

Starting from the 1970's and 1980's, it is known that "hot carrier effects" can cause long term reliability issues to MOS transistors.[70,71] A comparison of several approaches to suppress hot carrier effects in n-channel MOS transistors was discussed by Nayak *et al.*[72] The lightly doped drain (LDD) structure has been used to suppress hot carrier effects. For n-channel MOS transistors, there are two deep heavily doped n-type drain and source; in order to suppress hot carrier effects, two shallow lightly doped n-type regions known as NLDD are added. The NLDD region is adjacent to the deep heavily doped n-type draina and source and underneath the gate oxide. Similarly, for p-channel MOS transistors, there are two deep heavily doped p-type drain and source; in order to suppress hot carrier effects, two shallow lightly doped p-type regions known as PLDD are added. The PLDD region is adjacent to the deep heavily doped p-type draina and source and underneath the gate oxide.

In the beginning, the NLDD and PLDD regions were really "lightly doped". As scaling continues, the device dimensions become smaller and the supply voltage also becomes smaller. (Note: Lightly doped drain can help to suppress hot carrier effects. A smaller supply voltage can also help to suppress hot carrier effects.) For newer generations of CMOS technology, the NLDD and PLDD regions became "heavily doped" in order to have large drive current. The hot carrier effects may not be so serious because of the smaller supply voltage used for newer generations of CMOS.

The short channel effect (SCE) is the decrease of the magnitude of the threshold voltage when the gate length becomes smaller. An important approach is to use a thinner gate oxide to suppress the short channel effect. A smaller junction depth for the drain and source is also important. As shown in Fig. 2.2, the short channel effect is smaller for smaller junction depth x_j according to Miyake *et al.*[73]

In the early days, it was difficult to make very small energy ion implantation or very high energy ion implantation. Subsequently, very small energy ion implantation or very high energy ion implantation become possible. Ultra-shallow junction needs very small energy ion implantation. In addition, the advent of rapid thermal

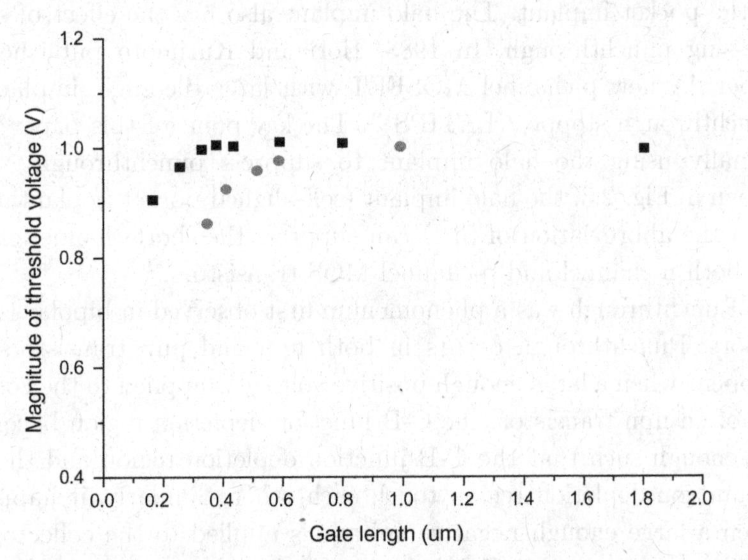

Fig. 2.2 Dependence of the magnitude of threshold voltage of p-channel MOSFET's on gate length for S-D junction-depth = 80 nm (squares) and for S-D junction-depth = 210 nm (circles). BF_2 channel implantation dose was 9×10^{12} cm^{-2}. (Modified from Fig. 9 in M. Miyake, T. Kobayashi and Y. Okazaki, "Subquarter-micrometer gate-length p-channel and n-channel MOSFETs with extremely shallow source-drain junctions", *IEEE Trans. Electron Dev.*, vol. 36, no. 2 (Feb. 1989), pp. 392–398.)

annealing (RTA) technology was also important for the development of ultra-shallow junction technology. For the implantation of n-type dopant for ultra-shallow junction, arsenic is usually preferred to phosphorus for silicon. (Note: Arsenic source can be a solid arsenic source or a gaseous source like arsine, AsH_3. Phosphorus source can be a solid phosphorus source or a gaseous source like phosphine, PH_3.) For the implantation of p-type dopant for ultra-shallow junction, B^+ ion implant or BF_2^+ ion implant are the usual choices. (Note: B^+ ions or BF_2^+ ions can be extracted from a gaseous source of boron trifluoride, BF_3.) In the early days, it was easier to achieve shallow junction for BF_2^+ ion implant compared to B^+ ion implant. The advent of very small energy ion implanter makes it possible to use B^+ ion implant for ultra-shallow junction formation.

Another important approach to suppress the short channel effect is the application of the halo implant. The halo implant is also known as the pocket implant. The halo implant also has the effect of suppressing punchthrough. In 1988, Hori and Kurimoto published a paper "A new p-channel MOSFET with large-tile-angle implanted punchthrough stopper (LATIPS)". The key point of this paper was actually using the halo implant to suppress punchthrough.[74] As shown in Fig. 2.3, the halo implant (self-aligned pocket implantation with the abbreviation of SPI) can suppress the short channel effect for both n-channel and p-channel MOS transistors.[75]

Punchthrough was a phenomenum first observed in bipolar transistors. Punchthrough occurs in both npn and pnp transistors. It happens when a large enough positive voltage is applied to the collector of an npn transistor; the C-B junction depletion region becomes big enough such that the C-B junction depletion region and the B-E junction depletion region touch each other. Similarly, it happens when a large enough negative voltage is applied to the collector of a pnp transistor; the C-B junction depletion region becomes big enough such that the C-B junction depletion region and the B-E junction depletion region touch each other. An n-channel enhancement mode MOS transistor has a built-in parasitic npn transistor; similarly, a p-channel enhancement mode MOS transistor has a built-in parasitic pnp transistor. To stop punchthrough in short MOS

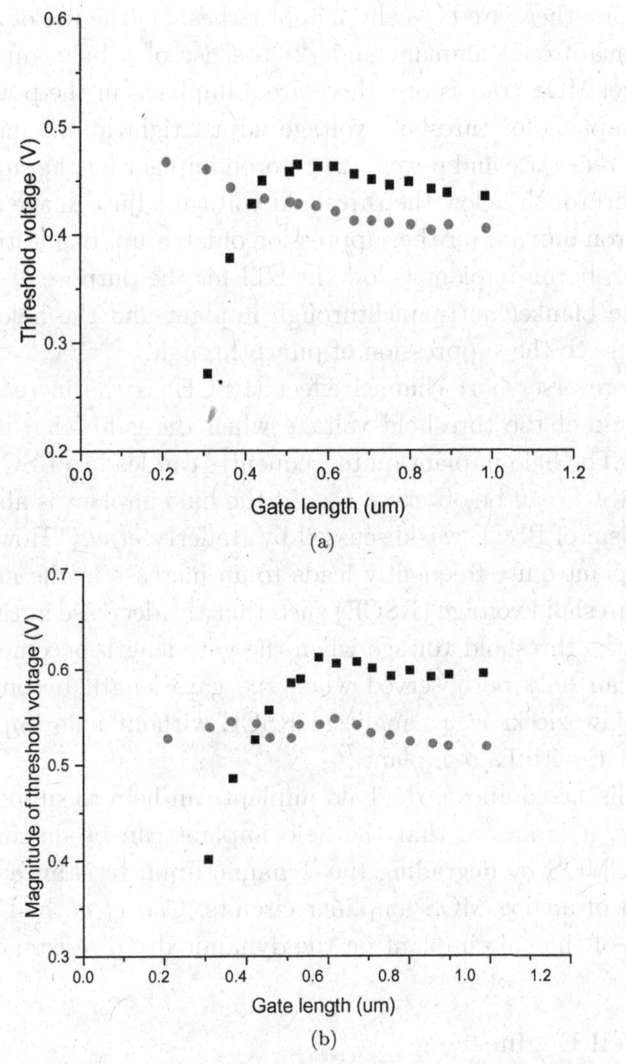

Fig. 2.3 Threshold voltage as a function of the gate length for SPI (self-aligned pocket implantation) and conventional LDD (lightly doped drain) devices (a) for n-channel MOS transistors and (b) for p-channel MOS transistors. (Note: Squares are for the case without pocket implant while circles are for the case with pocket implant.) This figure shows that the halo implant (also known as the pocket implant) can help to suppress the short channel effect for both n-channel and p-channel MOS transistors (Hori 1992[75]) (Modified from Fig. 3 in A. Hori, M. Segawa, H. Shimomura and S. Kameyama, "A self-aligned pocket implantation (SPI) technology for 0.2-μm dual-gate CMOS", *IEEE Electron Dev. Lett.*, vol. 13, no. 4 (April 1992), pp. 174–176.)

transistors, there are basically 2 approaches: (1) the use of a blanket anti-punchthrough implant and (2) the use of a halo implant. For n-channel MOS transistors, there are 3 implants in the p-well: (1) a boron implant for threshold voltage adjust right at the interface of the gate dielectric and p-well, (2) a boron implant for the suppression of punchthrough below the threshold voltage adjust image and (3) a deep boron implant for the suppression of latch-up. In addition, there may be a boron implant below the STI for the purpose of isolation. Both the blanket anti-punchthrough implant and the halo implant contribute to the suppression of punchthrough.

The reverse short channel effect (RSCE) is the increase of the magnitude of the threshold voltage when the gate length becomes smaller. The halo implant quite frequently can lead to RSCE. Sometimes, RSCE can be observed even if the halo implant is absent; one mechanism of RSCE was discussed by Rafferty *et al.*[76] However, the halo implant quite frequently leads to an increase in the magnitude of the threshold voltage (RSCE) such that the decrease in the magnitude of the threshold voltage when the gate length becomes smaller (SCE) can only be observed when the gate length becomes "very small". Gwoziecki *et al.* analyzed RSCE without halo implant and also RSCE with halo implant.[77]

As discussed above, the halo implant can help to suppress SCE. However, it is known that the halo implant can be detrimental to analog CMOS by degrading the dynamic drain resistance and thus the gain of analog MOS amplifier circuits. Cao *et al.* had made an analysis of the halo implant on the dynamic drain resistance.[78]

2.7 Well Engineering

For CMOS integrated circuits, it is possible to have p-well CMOS (starting from an n-type Si substrate) or n-well CMOS (starting from a p-type Si substrate). For p-well CMOS, n-channel MOS transistors will be degraded because of degraded electron mobility by additional impurities; for n-well CMOS, p-channel MOS transistors will be degraded because of degraded hole mobility by additional impurities. An example of p-well CMOS technology was discussed by Rung *et al.*[79] Similarly, an example of n-well CMOS technology was discussed by Martin and Chen.[80] For twin-well CMOS, there is a p-well and also an n-well; the starting substrate for twin-well CMOS can be n-type or p-type Si substrate). An example of twin-well CMOS technology was discussed by Stolmeijer.[81] After some years, twin-well CMOS technology is widely adopted.

Early CMOS technology used diffusion to form the p-well or n-well. Retrograde well technology involved the use of a high-density implanter to form the p-well or n-well. One of the advantages of retrograde well technology is the suppression of latch-up.[82] (Note: Latch-up is the phenomenon that a CMOS integrated circuit virtually becomes a short circuit because of some transient effect or bombardment by high energy radiation. Once this short circuit is formed, the only way to stop it is to turn off the power supply of the CMOS integrated circuit.) Retrograde well technology can be used for p-well, n-well or twin-well CMOS technology.

In the early days, it was difficult to make very small energy ion implantation or very high energy ion implantation. Subsequently, very small energy ion implantation or very high energy ion implantation become possible. Retrograde well technology needs very high energy ion implantation.

Figure 2.4 shows the increase of mobility with the increase of BF_2 channel implantation dose for p-channel MOSFET. (Note: This is the case for a buried-channel p-channel MOS transistor. This is old technology!)

There is a blanket anti-punchthrough implant in the p-well to suppress the punchthrough of the n-channel MOS transistor. Similarly, there is a blanket anti-punchthrough implant in the n-well to suppress the punchthrough of the p-channel MOS transistor. The design of the blanket anti-punchthrough implant can be considered part of the well engineering.

2.8 Technology After MOS Transistor Formation

After the adoption of polysilicon as the gate material, there was a problem that the resistivity of polysilicon is high. The next step

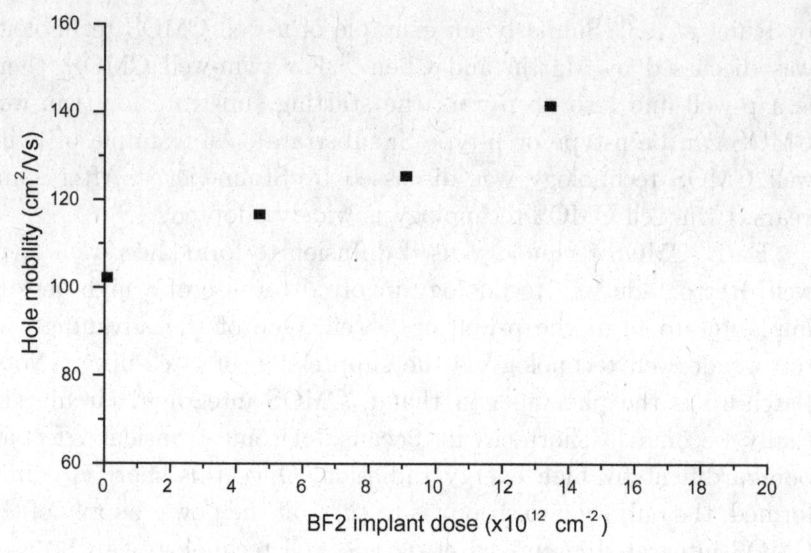

Fig. 2.4 Relationship between carrier mobility and BF_2 channel implantation dose for p-channel MOSFET'a with 80-nm S-D junctions. Hole mobility at the channel was estimated at $V_{GS} = V_{th}$ and $V_{DS} = -0.01$ V. (Modified from Fig. 12 in M. Miyake, T. Kobayashi and Y. Okazaki, "Subquarter-micrometer gate-length p-channel and n-channel MOSFETs with extremely shallow source-drain junctions", *IEEE Trans. Electron Dev.*, vol. 36, no. 2 (Feb. 1989), pp. 392–398.)

was to add a layer of silicide on top of the polysilicon to lower its resistivity. There was two main approaches of silicide technology. The first one was known as polycide technology; silicide was only applied to the gate. The second one was known as salicide (self-aligned silicide) technology; silicide was applied to the gate, the drain and the source. Quite frequently, the silicide for polycide technology was tungsten silicide. The silicide for silicide technology was titanium silicide for some time. Later, cobalt silicide replaced titanium silicide. Subsequently, nickel silicide replaced cobalt silicide. When the line width became smaller, cobalt silicide performed better than titanium silicide; when the line width became even smaller, nickel silicide performed better than cobalt silicide. When silicide technology is used, there is silicide on the drain and source, silicide junction spiking may cause serious leakage current if the drain to well or source to well junction is too shallow. If the junction is too deep, there will be serious "short channel effect". This contradiction is solved by using a combination of a deep drain/source junction to avoid junction spiking and an ultra-shallow drain/source extension near the channel region to avoid "short channel effect".

It turned out that the polycide technology was the preferred technology for DRAM (dynamic random access memory) while the salicide technology was the preferred technology for logic. The author believes that the polycide process is more compatible with the self-aligned contact (SAC) technology needed for DRAM. For DRAM technology, switching speed may not be so important but packing density can be important. There is some discussion on the polycide gate and SAC in the paper by Rupp *et al.*[83] For DRAM technology, the polycide technology migrated to poly-metal technology; poly-metal technology was basically a metal film (quite frequently tungsten) on top of polysilicon with a diffusion barrier to prevent silicidation.[84] The contact to the gate, the drain and the source was aluminum in the early days of CMOS technology. Subsequently, tungsten metal deposited by chemical vapor deposition (CVD) was used as the material for the contact to the gate, the drain and the source.

Aluminum was once the metal of choice for interconnection. The usual technique of deposition was sputtering. Dry etching was done on the metal. Inter-metal dielectric (IMD) was deposited to fill up the gap between metal lines. Subsequently, aluminum was replaced by copper. The usual technique of deposition was electroplating. Etching was done on the IMD before copper electroplating. Copper has higher conductivity compared to aluminum. In addition, copper performs better than aluminum in terms of electromigration. The IMD was initially silicon dioxide. The deposition technique was usually some sort of PECVD (plasma enhanced chemical vapor deposition). The melting point of aluminum is only about 600°C and so the IMD deposition cannot be a high temperature technique. (Note: The dielectric constant of silicon dioxide is 3.9. However, the dielectric constant of PECVD silicon dioxide can be 4.2.) Later, insulators with lower dielectric constant than silicon dioxide was developed. This was known as low-k dielectric technology. The first popular low-k dielectric was FSG (fluorosilicate glass). FSG was basically fluorine doped silicon dioxide. Fluorine doping of silicon dioxide ca lower the dielectric constant, for example, to 3.6. Subsequently, low-k dielectric with much lower dielectric constant was developed. Recently, through silicon via (TSV) became a hot R&D research topic. Copper electroplating technology is used for the fabrication of TSVs. Now, TSV technology has matured enough for manufacturing.

The technology of MOS transistor formation is quite frequently known as FEOL (front end of line) technology. The technology after MOS transistor formation is quite frequently known as BEOL (back end of line) technology. FEOL technology is, of course, important for high speed and low power. BEOL technology also can also influence the switching speed and power consumption. For manufacturing, BEOL technology is very important in terms of yield enhancement.

2.9 FinFET

Traditionally, MOS transistor technology has been based on the "planar technology" developed in the 1960's. Now in the 21st century, three-dimensional device structures have become possible. FinFET technology is a big breakthrough in MOS device technology. Conventional (100) silicon has notch in the ⟨110⟩ direction. Silicon has the same crystal structure as diamond. The best cleavage plane is the (111) family of planes. However, wafer dicing along the ⟨111⟩ family of directions will result in a triangular die. People prefer rectangular dies. The second best cleavage plane is the (110) family of planes. Wafer dicing along the ⟨110⟩ family of directions will result in a rectangular die. For FINFETs fabricated on (100) silicon with notch in the ⟨110⟩ direction, electron mobility is reduced while hole mobility is enhanced compared with conventional bulk technology. In addition, the subthreshold swing of FinFET is significantly improved over planar MOS transistors. Thus FinFET technology can improve (1) packing density, (2) switching speed and (3) power consumption. In addition, FINFET technology produces smaller threshold voltage mismatch and higher Early voltage; these advantages are important for analog CMOS technology. Historically, the on current of p-channel

MOS transistors is usually about one half of that of n-channel MOS transistors. For FINFET technology, the on current of p-channel MOS transistors is about the same as that of n-channel MOS transistors. More discussion on FINFETs can be found in Chapter Three, Chapter Four and Chapter Five.

In her 2002 IEEE Spectrum article, Linda Geppert described the FINFET structure which was studied in various R&D laboratories in the world at that time.[85] In their 2011 IEEE Spectrum article, Ahmed and Schuegraf mentioned that in May 2011 Intel had announced their plan to go for 3-D transistor structures in their future product.[86] That is Intel will adopt the FINFET structure in the near future. Thus a discussion on the FINFET structure is important. (Note: Intel called their technology "tri-gate" transistor technology.)

2.10 Reliability

The drift of sodium ions under an electric field can cause a shift of the threshold voltage. Threshold voltage instability due to sodium ions or other alkali ions was once an important issue in MOS technology. This problem can be solved by "clean" processing. In addition, FEOL reliability issues include (1) hot carrier effects, (2) NBTI (negative bias threshold instability) and PBTI (positive bias threshold instability).

Hot carrier effects in MOS transistors have been observed. In the relatively early days, hot carrier effects are stronger in n-channel MOS transistors. In more advanced CMOS technology, hot carrier effects in p-channel MOS transistors can also be strong.[87] A short discussion on the suppression of hot carrier effects was given in Section 2.6.

For silicon dioxide or silicon oxynitride based gate dielectric, NBTI is the threshold voltage instability of p-channel MOS transistors when a negative gate-to-source voltage is applied. For high-k based gate dielectric, PBTI is the threshold voltage instability of n-channel MOS transistors when a positive gate-to-source voltage is applied.[88]

Electromigration is the movement of metal in the presence of an electron wind. It can cause either an open circuit or short circuit. When aluminum based BEOL technology is used, copper was added to aluminum to suppress electromigration. Subsequently, aluminum based BEOL technology migrated to copper based BEOL technology. As discussion above, copper performs better compared to aluminum in terms of electromigration.

References

[1] J. E. Lilienfled, "Method and apparatus for controlling electric currents", US Patent 1,745,175, filed in 1926 and awarded in 1930.

[2] J. E. Lilienfled, "Device for controlling electric currents", US Patent 1,900,018, filed in 1928 and awarded in 1933.

[3] O. Heil, "Improvements in or relating to electrical amplifiers and other control arrangements and devices", British Patent 439,457, filed in 1935 and awarded in 1935.

[4] M. Riordan and L. Hoddeson, "Crystal Fire: The Invention of the Transistor and the Birth of the Information Age", Norton, New York, 1997.

[5] I. Tamm, "Uber eine mogliche art der elektronenbindung an kristalloberflachen", Phys. Z. Soviet Union, vol. 1 (1932), pp. 733–746. (Note: The title of the paper after translation from German to English is "On the possible bound states of electrons on a crystal surface". The full name of the journal is Physik Zeitschrift der Sowjetunion. This paper was written in German and has been re-printed in "I.E. Tamm Selected Works", edited by B.M. Bolotovskii and V. Ya. Frenkel, Springer-Verlag, Berlin 1991, pp. 92–102.)

[6] W. Shockley, "On the surface states associated with a periodic potential", *Phys. Rev.*, vol. 56, no. 4 (1939), pp. 317–323.

[7] W. Shockley, "A unipolar field-effect transistor", *Proc. IRE*, vol. 40, no. 11 (Nov. 1952), pp. 1365–1376.

[8] G. C. Dacey and I. M. Ross, "Unipolar field-effect transistor", *Proc. IRE*, vol. 41, no. 8 (Aug. 1953), pp. 970–979.

[9] C. A. Mead, "Schottky barrier gate field effect transistor", *Proc. IEEE*, vol. 54, no. 2 (Feb. 1966), pp. 307–309.

[10] M. M. Atalla, E. Tannenbaum and E. J. Scheibner, "Stabilization of silicon surfaces by thermally grown oxides", *Bell System Technical Journal*, vol. 38, no. 3 (May 1959), pp. 749–783.

[11] D. Kahng and M. M. Atalla, "Silicon-silicon dioxide field induced surface devices", IRE/AIEE Solid-State Device Research Conference, Carnegie Institute of Technology, Pittsburgh, Pennsylvania, USA, 1960.

[12] D. Kahng, "Silicon-silicon dioxide surface device", Technical memorandum of Bell Laboratories issued on January 16, 1961. This paper has been reprinted in the book *Semiconductor Devices: Pioneering Papers*, edited

by S. M. Sze, World Scientific, Singapore, 1991: D. Kahng, "Silicon-silicon dioxide surface device", pp. 583–596.

[13] D. Kahng, "Electric controlled semiconductor device", US Patent 3,102,230, filed in 1960 and awarded in 1963.

[14] D. Kahng, "A historical perspective on the development of MOS transistors and related devices", *IEEE Trans. Electron Dev.*, vol. 23, no. 7 (July 1976), pp. 655–657.

[15a] P. Balk, "Effects of hydrogen annealing on silicon surfaces", Electrochemical Society Spring Meeting, San Francisco, California, USA, 1965.

[15b] L. Do Thanh and P. Balk, "Elimination and generation of Si-SiO2 interface traps by low temperature hydrogen annealing", *J. Electrochem. Soc.*, vol. 135, no. 7 (July 1988), pp. 1797–1801.

[16] E. H. Nicollian and J. R. Brews, *MOS Physics and Technology*, Wiley, New York, 2003, pp. 1–906.

[17] R. G. Arns, "The other transistor: early history of the metal-oxide semiconductor field-effect transistor", *IEEE Engineering and Education Journal*, vol. 7, no. 5 (Oct. 1998), pp. 233–240.

[18] M. J. Riezenman, "Wanlass's CMOS circuit", *IEEE Spectrum*, vol. 28, no. 5 (May 1991), p. 44.

[19] R. K. Bassett, "MOS technology, 1963–1974: A dozen crucial years", *The Electrochemical Society Interface*, Fall 2007, pp. 46–50.

[20] B. Lojek, "History of semiconductor engineering", Springer, New York, 2007. (According to Lojek's book, it appears that Wanlass worked for some time on MOS transistor stability in Fairchild and then left. Then Snow joined Fairchild and published an important paper in JAP in 1965. Thus there is a possibility that Wanlass had done some pioneering work on MOS transistor stability in Fairchild and it was followed up by Snow. Snow did not mention the name of Wanlass in his work.)

[21] E. H. Snow, A. S. Grove, B. E. Deal and C. T. Sah, "Ion transport phenomena in insulating films", *J. Appl. Phys.*, vol. 36, no. 5 (May 1965), pp. 1664–1673.

[22] H. Schmid, "Making LSI circuits: a comparison of processing techniques", *IEEE Transactions on Manufacturing Technology*, vol. 1, no. 2 (Dec. 1972), pp. 19–31.

[23] D. Matsuura, H. Ozawa, M. Tohiguchi, M. Uchino, E. Miyata, H. Tsunemi, T. Inui, T. G. Tsuru, Y. Kamata, H. Nakaya, S. Miyazaki, K. Miyaguchi, M. Muramatsu, H. Suzuki and S. Takagi, "Development of p-channel charge-coupled device for NeXT, the next Japanese X-ray astronomical satellite mission", *Jpn. J. Appl. Phys.*, vol. 45, no. 11 (Nov. 2006), pp. 8904–8909.

[24] J. Borel, J. Bernard and J. P. Suat, "A depletion load self-aligned technology", *Solid-State Electron.*, vol. 16, no. 12 (Dec. 1973), pp. 1377–1381.

[25] B. Lojek, *History of Semiconductor Engineering*, Springer, New York, 2007.

[26] K. G. Aubuchon, "The use of ion implantation to set the threshold voltage of MOS transistors", Proceedings of the International Conference on Properties and Use of M.I.S. Structures, pp. 575–590 (1969).

[27] M. R. MacPherson, "The adjustment of MOS transistor threshold voltage by ion implantation", *Appl. Phys. Lett.*, vol. 18, no. 11, (1 June 1971) pp. 502–504.
[28] M. R. MacPherson, "Threshold shift calculations for ion implanted MOS devices", *Solid-State Electron.*, vol. 15, no. 12, (Dec. 1972) pp. 1319–1326.
[29] T. W. Sigmon and R. Swanson, "MOS threshold shifting by ion implantation", *Solid-State Electron.*, vol. 16, no. 11, (Nov. 1973) pp. 1217–1232.
[30] A. Y. Jaddam, "Fabrication of enhancement n-channel MOSFET's using ion implanted boron for controlled channel doping", *IEDM Abstracts*, (1971) p. 156.
[31] F. M. Wanlass and C. T. Sah, "Nanowatt logic using using field effect metal-oxide semiconductor triodes", IEEE International Solid-State Circuits Conference (ISSCC), Digest of Technical Papers, pp. 32–33, Feb. 1963. This paper has been reprinted in the book *Semiconductor Devices: Pioneering Papers*, edited by S. M. Sze, World Scientific, Singapore, 1991: F. M. Wanlass and C. T. Sah, "Nanowatt logic using field effect metal-oxide semiconductor triodes", pp. 637–638.
[32] F. M. Wanlass, "Low stand-by power complementary field effect circuitry", US Patent 3,356,858, filed in 1963 and awarded in 1967.
[33] J. M. Miller, "Dependence of input impedance of a three-electrode vacuum tube upon the load in the plate circuit," *Scientific Papers of the National Bureau of Standards*, vol. 15, no. 351 (1919–1920), pp. 367–385.
[34] J. Millman and C. C. Halkias, *Integrated Electronics: Analog and Digital Circuits and Systems*, McGraw-Hill, New York, 1972, pp. 255–256.
[35] R. W. Bower, H. G. Dill, K. G. Aubuchon and S. A. Thompson, "MOS field effect transistors formed by gate masked ion implantation", *IEEE Trans. Electron Dev.*, vol. 15, no. 10 (Oct. 1968), pp. 757–761.
[36] R. E. Kerwin, D. L. Klein and J. C. Sarace, "Method for making MIS structures", US Patent 3,475,234, filed in 1967 and awarded in 1969.
[37] J. C. Sarace, R. E. Kerwin, D. E. Klein and R. Edwards, "Metal-nitride-oxide-silicon field-effect transistors, with self-aligned gates", *Solid-State Electron.*, vol. 11, no. 7 (July 1968), pp. 653–660.
[38] L. L. Vadasz, A. S. Grove, T. A. Rowe and G. E. Moore, "Silicon-gate technology", *IEEE Spectrum*, vol. 6, no. 10 (Oct. 1969), pp. 28–35.
[39] F. Faggin and T. Klein, "Silicon gate technology", *Solid-State Electron.*, vol. 13, no. 8 (Aug. 1970), pp. 1125–1144.
[40] D. H. Lee and J. W. Mayer, "Ion-implanted semiconductor devices", *Proc. IEEE*, vol. 62, no. 9 (Sep. 1974), pp. 1241–1255.
[41] F. Faggin, M. E. Hoff Jr., S. Mazor and M. Shima, "The history of the 4004", *IEEE Micro*, vol. 16, no. 6 (Dec. 1996), pp. 10–20.
[42] F. Faggin, "The making of the first microprocessor", *IEEE Solid-State Circuits Magazine*, vol. 1, no. 1 (Winter 2009), pp. 8–21.
[43] K. M. Cham and S.-Y. Chiang, "Device design for the submicrometer p-channel FET with n^+ polysilicon gate", *IEEE Trans. Electron Dev.*, vol. 31, no. 7 (July 1984), pp. 964–968.

[44] A. Hori, M. Segawa, H. Shimomura and S. Kameyama, "A self-aligned pocket implantation (SPI) technology for 0.2-μm dual-gate CMOS", *IEEE Electron Dev. Lett.*, vol. 13, no. 4 (April 1992), pp. 174–176.

[45] A. Hori, M. Segawa, S. Kameyama and M. Yaushira, "High-performance dual-gate CMOS utilizing a novel self-aligned pocket implantation (SPI) technology", *IEEE Trans. Electron Dev.*, vol. 40, no. 9 (Sept. 1993), pp. 1675–1681.

[46] K. M. Cham, D. W. Wenocur, J. Lin, C. K. Lau and H.-S. Fu, "Submicrometer thin gate oxide p-channel transistors with p^+ polysilicon gates for VLSI applications", *IEEE Electron Dev. Lett.*, vol. 7, no. 1 (Jan. 1986), pp. 49–52.

[47] G. J. Hu and R. H. Bruce, "Design tradeoffs between surface and buried-channel FET's", *IEEE Trans. Electron Dev.*, vol. 32, no. 3 (March 1985), pp. 584–588.

[48] Z. J. Ma, J. C. Chen, Z. H. Liu, J. T. Krick, Y. C. Cheng, C. Hu and P. K. Ko, "Suppression of boron penetration in p^+ polysilicon gate p-MOSFET's using low-temperature gate-oxide N_2O anneal", *IEEE Electron Dev. Lett.*, vol. 15, no. 3 (March 1994), pp. 109–111.

[49] L. K. Han, D. Wristers, J. Yan, M. Bhat and D. L. Kwang, "Highly suppressed boron penetration in NO-nitrided SiO_2 for p^+-polysilicon gated MOS device applications", *IEEE Electron Dev. Lett.*, vol. 16, no. 7 (July 1995), pp. 319–321.

[50] G. Yaron and D. Frohman-Bentchkowsky, "Capacitance voltage characterization of poly Si-SiO_2-Si structures", *Solid-State Electron.*, vol. 23, no. 5 (May 1980), pp. 433–439.

[51] C.-Y. Lu, J. M. Sung, H. C. Kirsch, S. J. Hillenius, T. E. Smith and L. Manchanda, "Anomalous C-V characteristics of implanted poly MOS structure in n^+/p^+ dual-gate CMOS technology", *IEEE Electron Dev. Lett.*, vol. 10, no. 5 (May 1989), pp. 192–194.

[52] Y. Ohkura, "Quantum effects in Si n-MOS inversion layer at high substrate concentration", *Solid-State Electron.*, vol. 33, no. 12 (Dec. 1990), pp. 1581–1585.

[53] S.-H. Lo, D. A. Buchanan and Y. Taur, "Modeling and characterization of quantization, polysilicon depletion, and direct tunneling effects in MOSFETs with ultrathin oxides", *IBM J. Res. Develop.*, vol. 43, no. 3 (May 1999), pp. 327–337.

[54] H. Watanabe, "Depletion layer of gate poly-Si," *IEEE Trans. Electron Dev.*, vol. 52, no. 10 (Oct. 2005), pp. 2265–2271.

[55] T.-J. King, J. P. McVittie, K. C. Saraswat, and J. R. Pfiester, "Electrical properties of heavily doped polycrystalline silicon-germanium films," *IEEE Trans. Electron Dev.*, vol. 41, no. 2 (Feb. 1994), pp. 228–232.

[56] Y. V. Ponomarev, P. A. Stolk, C. Salm, J. Schmitz and P. H. Woerlee, "High-performance deep submicron CMOS technologies with polycrystalline-SiGe gates", *IEEE Trans. Electron Dev.*, vol. 47, no. 4 (April 2000), pp. 848–855.

[57] S. Inaba, R. Katsumata, H. Akatsu, R. Rengarajan, P. Ronsheim, C. S. Murphy, K. Sunouchi and G. B. Bronner, "Threshold voltage roll-up/roll-off characteristic control in sub-0.2-μm single workfunction gate CMOS for high-performance DRAM applications", *IEEE Trans. Electron Dev.*, vol. 49, no. 2 (Feb. 2002), pp. 308–313.

[58] J.-H. Sim and K. Kim, "Source-bias dependent charge accumulation in p^+-poly gate SOI dynamic random access memory cell transistors", *Jpn. J. Appl. Phys.*, vol. 37, Part 1, no. 3B (March 1998), pp. 1260–1263.

[59] Y. Hiura, A. Azuma, K. Nakajima, Y. Akasaka, K. Miyano, H. Nitta, A. Honjo, K. Tsuchida, Y. Toyoshima, K. Suguro and Y. Kohyama, "Integration technology of polymetal (W/WSiN/poly-Si) dual gate CMOS for 1Gbit DRAMs and beyond ", *IEDM Tech. Dig.*, (1998), pp. 389–392.

[60] M. T. Bohr, R. S. Chau, T. Ghani and K. Mistry, "The high-k solution", *IEEE Spectrum*, vol. 44, no. 10 (Oct. 2007), pp. 29–35.

[61] J. N. A. Matthews, "Semiconductor industry switches to hafnium-based transistors", *Physics Today*, vol. 61, no. 2 (Feb. 2008), pp. 25–26.

[62] D. G. Schlom, S. Guha and S. Datta, "Gate oxides beyond SiO_2", *MRS Bulletin*, vol. 33, no. 11 (Nov. 2008), pp. 1017–1025.

[63] S. Mayuzumi, S. Yamakawa, Y. Tateshita, T. Hirano, M. Nakata, S. Yamaguchi, K. Tai, H. Wakabayashi, M. Tsukamoto, and N. Nagashima, "High-performance metal/high-k n- and p-MOSFETs with top-cut dual stress liners using gate-last damascene process on (100) substrates", *IEEE Trans. Electron Dev.*, vol. 56, no. 4 (Apr. 2009), pp. 620–626.

[64] E. Kooi, "Methods of producing a semiconductor device and a semiconductor device produced by said method", US Patent 3,970,486, filed in 1975 and awarded in 1976.

[65] E. Kooi, "The history of LOCOS", *IEEE*, New York, 1991, pp. 1–163.

[66] G. Ning, P.-C. Lin, C. Xing, A. Bian, H.-B. Zhao and Y.-L. Cao, "A robust shallow trench isolation high density plasma chemical vapor deposition void free process for 0.13μm CMOS technology", *ECS Transactions*, vol. 34, no. 1 (2011), pp. 743–748.

[67] A. Tavernier, L. Favennec, T. Chevolleau and V. Jousseaume, "Innovative gap-fill strategy for 28 nm shallow trench isolation", *ECS Transactions*, vol. 45, no. 3 (2012), pp. 225–232.

[68] W. S. Lau, K. S. See, C. W. Eng, W. K. Aw, K. H. Jo, K. C. J. Y. M. Lee, E. K. B. Quek, H. S. Kim, S. T. H. Chan and L. Chan, "Anomalous narrow width effect in NMOS and PMOS surface channel transistors using shallow trench isolation", *Proc. IEEE EDSSC 2005*, pp. 773–776.

[69] W. S. Lau, K. S. See, C. W. Eng, W. K. Aw, K. H. Jo, K. C. J. Y. M. Lee, E. K. B. Quek, H. S. Kim, S. T. H. Chan and L. Chan, "Anomalous narrow width effect in p-channel metal-oxide-semiconductor surface channel transistors using shallow trench isolation technology", *Microelectronics Reliability*, vol. 48, no. 6 (June 2008), pp. 919–922.

[70] S. A. Abbas and R. C. Dockerty, "Hot-carrier instability in IGFET's", *Appl. Phys. Lett.*, vol. 27, no. 3 (1 August 1975), pp. 147–148.

[71] K. K. Ng and G. W. Taylor, "Effects of hot-carrier trapping in n- and p-channel MOSFET's", *IEEE Trans. Electron Dev.*, vol. 30, no. 8 (Aug. 1983), pp. 871–876.

[72] D. K. Nayak, M.-Y. Hao, J. Umali and R. Rakkhit, "A comprehensive study of performance and reliability of P, As and hybrid As/P nLDD junctions for deep-submicron CMOS logic technology", *IEEE Electron Dev. Lett.*, vol. 18, no. 6 (June 1997), pp. 281–283.

[73] M. Miyake, T. Kobayashi and Y. Okazaki, "Subquarter-micrometer gate-length p-channel and n-channel MOSFETs with extremely shallow source-drain junctions", *IEEE Trans. Electron Dev.*, vol. 36, no. 2 (Feb. 1989), pp. 392–398.

[74] T. Hori and K. Kurimoto, "A new p-channel MOSFET with large-tile-angle implanted punchthrough stopper (LATIPS)", *IEEE Electron Dev. Lett.*, vol. 9, no. 12 (December 1988), pp. 641–643.

[75] A. Hori, M. Segawa, H. Shimomura and S. Kameyama, "A self-aligned pocket implantation (SPI) technology for 0.2-μm dual-gate CMOS", *IEEE Electron Dev. Lett.*, vol. 13, no. 4 (April 1992), pp. 174–643.

[76] C. S. Rafferty, H.-H. Vuong, S. A. Eshraghi, M. D. Giles, M. R. Pinto and S. J. Hillenius, "Explanation of reverse short channel effect by defect gradients", *IEDM Tech. Dig.*, pp. 311–314 (1993).

[77] R. Gwoziecki, T. Skotnicki, P. Bouillon and P. Gentil, "Optimization of V_{th} roll-off in MOSFET's with advanced channel architecture — Retrograde doping and pockets", *IEEE Trans. Electron Dev.*, vol. 46, no. 7 (July 1999), pp. 1551–1561.

[78] K. M. Cao, W. Liu, X. Jin, K. Vasanth, K. Green, J. Krick, T. Vrotsos and C. Hu, "Modeling of pocket implanted MOSFETs for anomalous analog behavior", *IEDM Tech. Dig.*, pp. 171–174 (1999).

[79] R. D. Rung, C. J. Dell'oca and L. G. Walker, "A retrograde p-well for higher density CMOS", *IEEE Trans. Electron Dev.*, vol. 28, no. 10 (October 1984), pp. 1115–1119.

[80] R. A. Martin and J. Y.-T. Chen, "Optimized retrograde n-well for 1-μm CMOS technology", *IEEE Journal of Solid-State Circuits*, vol. 21, no. 2 (April 1986), pp. 286–292.

[81] A. Stolmeijer, "A twin-well CMOS process employing high-energy ion implantation", *IEEE Trans. Electron Dev.*, vol. 33, no. 4 (April 1986), pp. 450–457.

[82] A. G. Lewis, R. A. Martin, T.-Y. Huang, J. Y. Chen and M. Koyanagi, "Latchup performance of retrograde and conventional n-well CMOS technologies", *IEEE Trans. Electron Dev.*, vol. 34, no. 10 (Oct. 1987), pp. 2156–2164.

[83] T. S. Rupp, D. Dobuzinsky, Z. Lu, V. J. Sardesai, H.-Y. Liu, M. Maldei, J. Faltermeier and J. Gambino, "High yielding self-aligned contact process for a 0.150-μm DRAM technology", *IEEE Trans. Semiconductor Manufacturing*, vol. 15, no. 2 (May 2002), pp. 223–228.

[84] R. Rengarajan, B. He, C. Ransom, J. C. Chang, R. Ramachandran, H. Yang, S. Butt, S. Halle, W. Yan, K. Lee, M. Chudzik, W. Robl, C. Parks,

J. G. Massey, G. La Rosa, Y. Li, C. Radens, R. Divakaruni and E. Crabbe, "1.5-V single work-function W/WN/n+-poly gate CMOS device design with 110-nm buried-channel PMOS for 90-nm vertical-cell DRAM", *IEEE Electron Dev. Lett.*, vol. 23, no. 10 (Oct. 2002), pp. 621–623.

[85] L. Geppert, "The amazing vanishing transistor act", *IEEE Spectrum*, vol. 39, no. 10 (Oct. 2002), pp. 50–53 and pp. 28–33.

[86] K. Ahmed and K. Schuegraf, "Transistor wars: rival architectures face off in a bid to keep Moore's law alive", *IEEE Spectrum*, vol. 48, no. 11 (Nov. 2011), pp. 50–53 and pp. 63–66.

[87] S.-Y. Chen, C.-H. Tu, J.-C. Lin, M.-C. Wang, P.-W. Kao, M.-H. Lin, S.-H. Wu, Z.-W. Jhou, S. Chou, J. Ko and H.-S. Haung, "Investigation of DC hot-carrier degradation at elevated temperatures for p-channel metal-oxide-semiconductor field-effect transistors of 0.13 μm technology", *Jpn. J. Appl. Phys.*, vol. 47, no. 3 (March 2008), pp. 1527–1531.

[88] F. Crupi, C. Pace, G. Cocorullo, G. Groeseneken, M. Aoulaiche and M. Houssa, "Positive bias temperature instability in nMOSFETs with ultra-thin Hf-silicate gate dielectrics", *Microelectronics Reliability*, vol. 80 (17 June 2005), pp. 130–133.

Chapter Three

CMOS Switching Speed Characterization and An Overview Regarding How to Speed Up CMOS

3.1 Introduction of Switching Speed

Switching speed is important for digital integrated circuits. This is the case no matter whether the fabrication technology is PMOS, NMOS or CMOS. In 1969, Vadasz *et al.* published an article about silicon gate PMOS technology.[1] They pointed out that the characteristic switching time constant of an inverter is given by $R_{eq} \times C_{load}$, where R_{eq} is a nonlinear resistance and C_{load} is the capacitance at the load. It is not too difficult to imagine that the drive current I_{drive} is inversely proportional to R_{eq}. Thus, we can say that the switching speed is roughly proportional to the drive current I_{drive} and inversely proportional to the load capacitance C_{load}. Thus the key concept to improve switching speed is to increase the drive current and to decrease the parasitic load capacitances. This concept is true no matter whether the fabrication technology is PMOS, NMOS or CMOS. In this chapter, the author will discuss the methodology to characterize switching speed and then various methods to improve CMOS switching speed.

3.2 Measurement of Switching Speed

3.2.1 *Method 1: Ring oscillator method*

A ring oscillator is basically a phase shift oscillator using an odd number of inverters cascaded together. The oscillation frequency of a ring oscillator, f_{RO}, is given by

$$f_{RO} = 1/(NT_{inv\_delay}) \tag{3.1}$$

In Eq. (3.1), N is the number of inverters cascaded together and T_{inv_delay} is the delay time of an individual inverter. When the number of inverters is an even number, the output of the ring oscillator will end up in one of the two possible stable states. For example, when the number of inverters is "two", the ring oscillator will have two stable states as shown in Fig. 3.1.

Shohno *et al.* published a paper on a ring oscillator fabricated by PMOS technology in 1978.[2] Yoshimi *et al.* reported their application of the ring oscillator method to study the switching speed of 0.5 μm CMOS in 1987.[3] Right now, the ring oscillator method is a well established method to measure CMOS switching speed. Figure 3.2 shows results measured by 125-stage CMOS ring oscillator plotted against supply voltage according to Tsui *et al.* 1994.[4] More discussion on the use of ring oscillators for process tuning has been provided, for example, by Bhusan *et al.*[5]

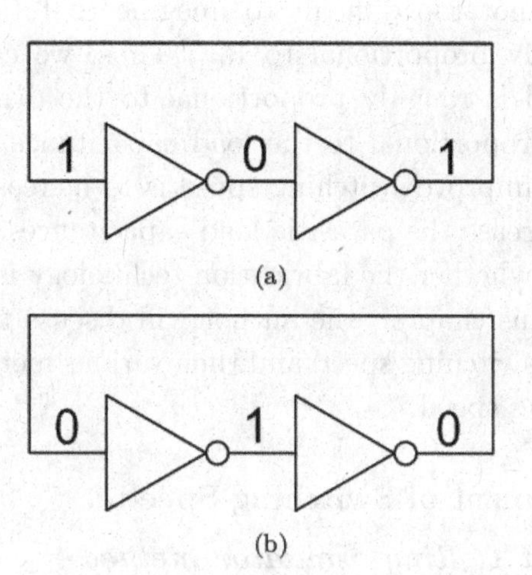

Fig. 3.1 When the number of inverters in a ring oscillator is an even number, for example 2, the ring oscillator will end up in either one of the two stable states (a) or (b) such that there is no oscillation. The number of inverters in a ring oscillator has to be an odd number in order for the ring oscillator to function.

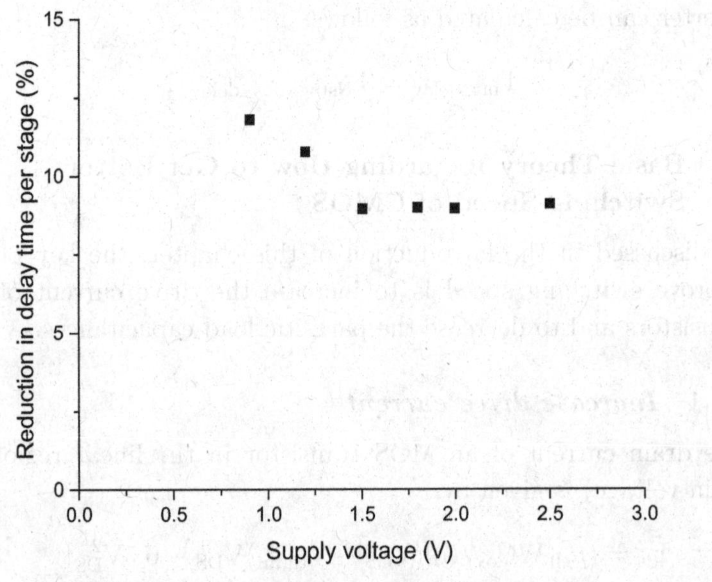

Fig. 3.2 Percentage improvement in the propagation delay time per stage measured by the ring oscillator technique due to the use of the oxynitride reoxidation process over that of the standard O_2 process at low voltage operating condition, according to Tsui *et al.*[4] (Modified from Fig. 9 in P. G. Y. Tsui, H.-H. Tseng, M. Orlowski, S.-W. Sun, P. J. Tobin, K. Reed and W. J. Taylor, "Suppression of MOSFET reverse short channel effect by N_2O gate poly reoxidation process", *IEDM Technical Digest*, (1994), pp. 501–504.)

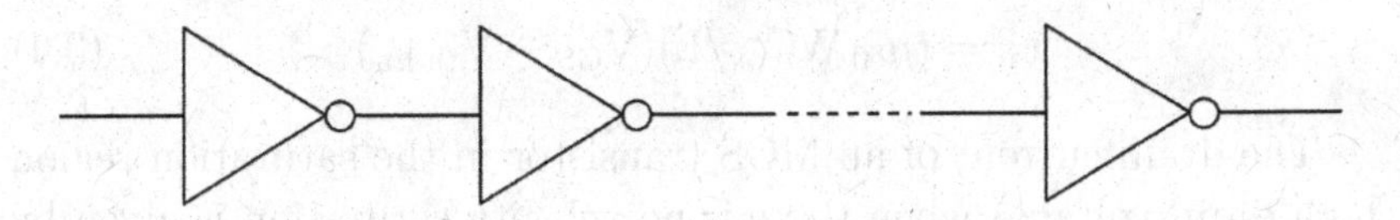

Fig. 3.3 The application of many CMOS inverters cascaded together as a test structure to measure propagation delay.

3.2.2 *Method 2: Cascading many logic gates together*

Figure 3.3 shows that many CMOS inverters can be cascaded together as a test structure to measure propagation delay. This approach has been discussed, for example, by Boeuf *et al.*[6] The propagation delay of N stages of inverters ($T_{Nstage\_inv\_delay}$) cascaded together can be measured and the propagation delay per stage of inverter can be calculated as follows:

$$T_{inv\_delay} = T_{Nstage\_inv\_delay}/N \quad (3.2)$$

3.3 Basic Theory Regarding How to Get Better Switching Speed of CMOS

As discussed in the introduction of this chapter, the key point to improve switching speed is to increase the drive current of MOS transistors and to decrease the parasitic load capacitances.

3.3.1 *Increase drive current*

The drain current of an MOS transistor in the linear region (low drain voltage) is given by

$$I_{ds} = (\mu_{eff} WC_{ox}/L)[(V_{GS} - V_{th,lin})V_{DS} - 0.5V_{DS}^2] \quad (3.3)$$

where μ_{eff} is the low-field effective mobility. W and L are the gate width and gate length, respectively. C_{ox} is the gate oxide capacitance per unit area. V_{GS} is the gate-to-source voltage. V_{DS} is the drain-to-source voltage. $V_{th,lin}$ is the linear threshold voltage (threshold voltage at low drain voltage).

and

$$g_m = (\mu_{eff} WC_{ox}/L)(V_{GS} - V_{th,lin}) \quad (3.4)$$

The drain current of an MOS transistor in the saturation region (high drain voltage) when there is no velocity saturation is given by

$$I_{ds} = [\mu_{eff} WC_{ox}/(2L)](V_{GS} - V_{th,sat})^2 \quad (3.5)$$

$$g_m = (\mu_{eff} WC_{ox}/L)(V_{GS} - V_{th,sat}) \quad (3.6)$$

The drain current of an MOS transistor in the saturation region (high drain voltage) when there is velocity saturation is given by

$$I_{ds} = (v_{sat} WC_{ox})(V_{GS} - V_{th,sat}) \quad (3.7)$$

$$g_m = v_{sat} WC_{ox} \quad (3.8)$$

$V_{th,sat}$ is the saturation threshold voltage (threshold voltage at high drain voltage) while $V_{th,lin}$ is the linear threshold voltage (threshold voltage at small drain voltage). In general, the saturation threshold voltage is smaller than the linear threshold voltage because of an effect known as "drain induced barrier lowering" (DIBL). Note: DIBL is quantitatively defined as the difference in the threshold voltages for a small drain voltage and for a large drain voltage respectively. Another approach is that DIBL is quantitatively defined as the difference in the threshold voltages for a small drain voltage and for a large drain voltage respectively divided by the difference in the drain voltages. For example, for an n-channel MOS transistor, the threshold voltage is +0.3 V at a drain voltage of 0.05 V; when the drain voltage is increased to 1 V, the threshold voltage is decreased to +0.2 V. Then DIBL according to the second way of definition is $(0.3 - 0.2)/(1 - 0.05) = 0.105\,\text{V/V}$. Similarly, for example, for a p-channel MOS transistor, the threshold voltage is −0.3 V at a drain voltage of −0.05 V; when the drain voltage is increased in the negative direction to −1 V, the threshold voltage is decreased in the negative direction to −0.2 V. The DIBL according to the second definition is $(0.3 - 0.2)/(1 - 0.05) = 0.105\,\text{V/V}$.

By inspecting Eq. (3.5) and Eq. (3.7), it can be easily seen that the drive current can be improved by increasing C_{ox}, for example, by using thinner gate oxide. This is true from the earliest days of MOS transistors down to the present state-of-the-art CMOS technology.

By inspecting Eq. (3.5), we can see that the drive current can be improved by reducing the channel length; however, by inspecting Eq. (3.7), we can see that the drive current cannot be improved by reducing the channel length. In reality, the MOS transistor operating in the saturation equation is not represented by Eq. (3.5) or Eq. (3.7). From the earliest days of MOS transistors down to the present state-of-the-art (45 nm CMOS), the drive current is still improved by reducing the channel length; however, the improvement is not as great as that predicted by Eq. (3.5).

By inspecting Eq. (3.5), we can see that the drive current can be improved by improving the carrier mobility; however, by inspecting Eq. (3.7), we can see that the drive current cannot be improved by improving the mobility. In reality, the MOS transistor operating in the saturation equation is not represented by Eq. (3.5) or Eq. (3.7). From the earliest days of MOS transistors down to the present state-of-the-art (45 nm CMOS), the drive current is still improved by improving the mobility; however, the improvement is not as great as that predicted by Eq. (3.5).

There is an important effect on the drive current due to series resistance. This cannot be seen in any of the equations above. The series resistance can be reduced, for example, by using silicide technology, etc. The series resistance can also be reduced, for example, by migrating from Al based back-end technology to Cu based back-end technology, etc.

Thus the drive current of MOS transistors can be improved by

(1) Increase C_{ox}, for example, by using thinner gate oxide or by using higher dielectric constant gate dielectric
(2) Reduce L, for example, by using smaller length gate electrode
(3) Improve mobility, for example, by stress engineering
(4) Reduce series resistance, for example, by using silicide technology in the front-end and by switching from aluminum based back-end to copper-based back-end technology

Conventional carrier transport theory has a low field mobility and also a high field saturation velocity as two important parameters to describe carrier transport. An important question is that: Are the low field mobility and the high field saturation velocity two independent quantities? Thornber[7] believed that low field mobility and the high field saturation velocity are two independent quantities. It appears to the author that the concept of velocity saturation was first proposed by some Bell Laboratories scientists (for example, E. J. Ryder and W. Shockley) in the early 1950's for the transport of electrons in germanium.[8,9] Subsequently, the concept of velocity saturation has been extended to silicon.[10]

The concept of "velocity saturation" is challenged and a phenomenon known as "velocity overshoot" was observed. Some scientists, for example Ruch,[11] pointed out that it needs some minimum distance for the drift velocity of electrons or holes to become a constant; for an MOS transistor with very short channel, the drift velocity of electrons or holes may become slightly larger than the saturation velocity of electrons or holes. This phenonmenon is known as "velocity overshoot". In 1985, Chou *et al.*[12] claimed that they had experimentally observed electron "velocity overshoot" in sub-100 nm MOS transistors.

In a 2009 paper, Lau *et al.*[13] pointed out that the drain current saturation in state-of-the-art short MOS transistors is due to "pinch off" instead of velocity saturation. Thus "velocity saturation" in the traditional sense may not exist in short MOS transistors. However, we can define an "effective saturation velocity". It turns out that the experimentally measured "effective saturation velocity" is not far away from the known value of "saturation velocity". Depending on the exact way of device fabrication, the effective saturation velocity can be slightly below or slightly above the known value of "saturation velocity". As shown in Fig. 3.4(a), the drain current I_{ds} versus drain voltage V_{DS} characteristics will show off some sort of saturation for both positive and negative drain voltage if there is "velocity saturation". As shown in Fig. 3.4(b), the drain current I_{ds} versus drain voltage V_{DS} characteristics will show off some sort of saturation for positive drain voltage but not for negative drain voltage if pinch-off is the dominant mechanism. As shown in Fig. 3.4(c), the drain current I_{ds} versus drain voltage V_{DS} characteristics will not show off any saturation for both positive and negative drain if there is no velocity saturation and pinch-off. Figure 3.5 shows actual experimental data similar to Fig. 3.4(b) and so it appeared that the drain current saturation is more likely to be due to pinch off instead of velocity saturation.

Device scaling implies that the channel length of MOS transistors can now be smaller than 0.1 μm. People have been wondering whether there is some sort of new physics for very short MOS transistors. In 1994, Kenji Natori proposed his ballistic MOS transistor theory.[14] Ballistic transport means that, without scattering, electrons or holes move according to Newton's second law of motion. According to the opinion of the author, a practical MOS transistor has to be heavily doped for threshold voltage control and suppression of

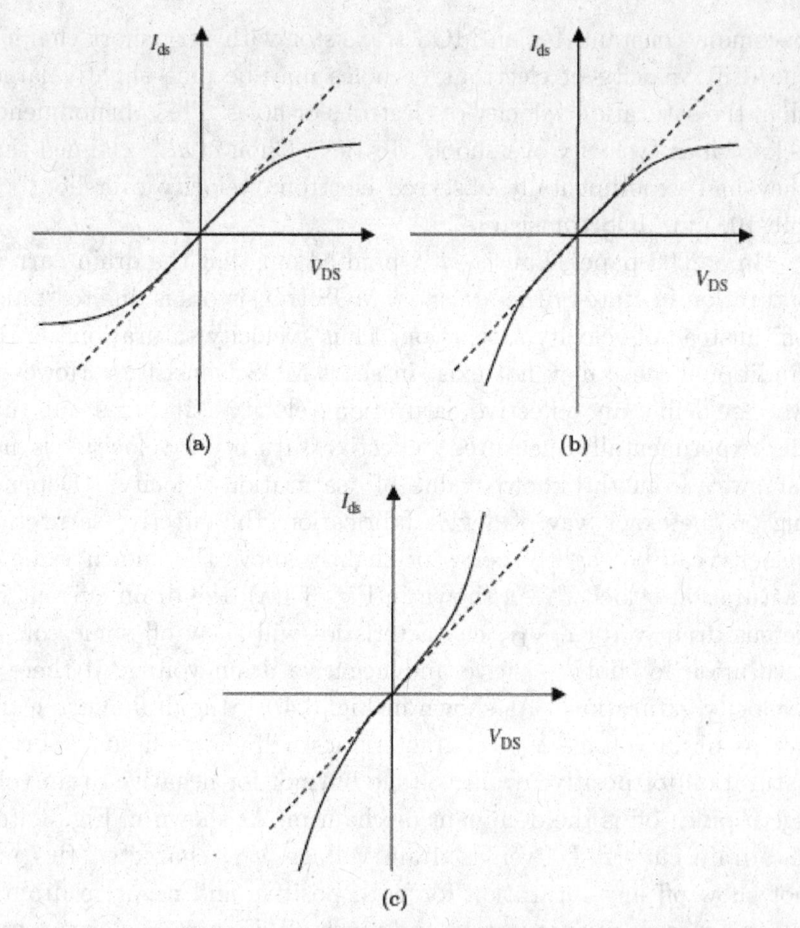

Fig. 3.4 Schematics of I_{ds} versus V_{DS} characteristics of an n-channel MOS transistor if drain current saturation is caused (a) by velocity saturation, (b) by pinch off and (c) if DIBL is dominant over velocity saturation or pinch off and there is no drain current saturation. This figure is similar to Fig. 6 from Lau *et al.* 2009.[13] (This figure comes from the work of the author.)

punchthrough such that ballistic transport is not likely to happen in practical devices. As shown in Fig. 3.6, the drain current is larger at 300 K compared to 77 K according to this ballistic transport.

However, experimentally, the drain current is usually found to be smaller for higher temperature. This is usually explained by a lower carrier mobility at higher temperature. For example, Fig. 3.7 shows

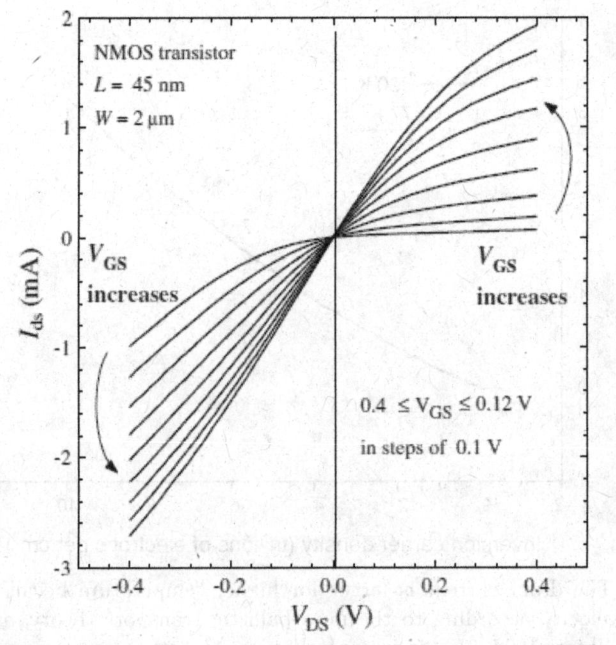

Fig. 3.5 Effects of negative V_{DS} and positive V_{DS} on I_{ds} versus V_{DS} characteristics of an n-channel MOS transistor (nominal gate length L = 45 nm). This figure is similar to Fig. 7 from Lau *et al.* 2009.[13] It appears to the author that drain current saturation is more likely to be due to pinch off instead of velocity saturation for state-of-the-art MOS transistors. (This figure comes from the work of the author.)

experimental data from Yang *et al.*[15]; it is obvious that the drain current is smaller at 125°C compared to −25°C.

A theory known as "quasi-ballistic transport" has also been proposed, for example, by Lundstrom.[16] This theory points out that the drain current is controlled by "mobility" instead of "saturation velocity" for both the linear region (low drain voltage) and also the saturation region (high drain voltage). For engineers, the key point of this "quasi-ballistic transport" is that higher mobility will improve the drain current for both the linear region (low drain voltage) and also the saturation region (high drain voltage). Thus in order to achieve better performance, the engineers should look for "methods" to improve the mobility.

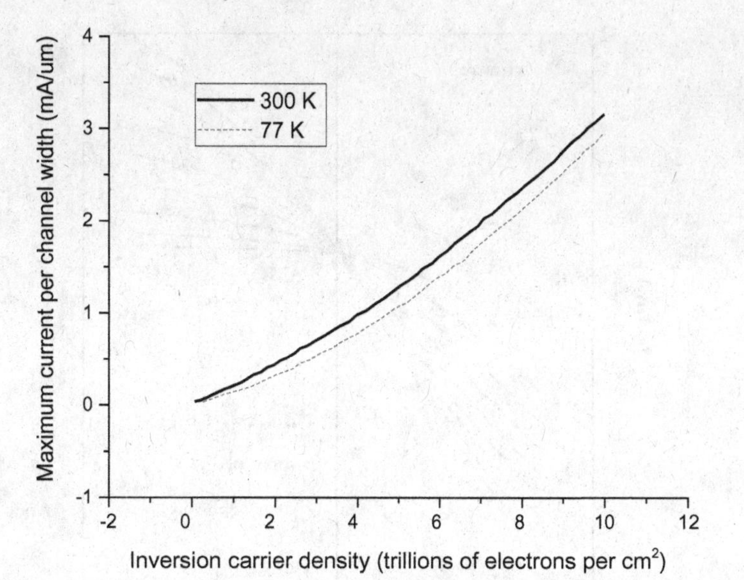

Fig. 3.6 The drain current is larger for higher temperature because of larger injection velocity according to the pure ballistic transport theory, as shown in Fig. 8 in Natori 1994.[14] This is contradictory to actual experimental observation; experimental data show that the drain current is smaller for higher temperature. The explanation is that the carrier mobility tends to be smaller at higher temperature. Thus the pure ballistic transport theory is not likely to be true in practical MOS transistors. (Modified from Fig. 8 in K. Natori, "Ballistic metal-oxide-semiconductor field effect transistor", *J. Appl. Phys.*, vol. 76, no. 8 (15 October 1994), pp. 4879–4890.)

The opinion of the author regarding the older theory (based on low field mobility and high field saturation velocity) and the newer quasi-ballistic theory that carriers may have only a few scattering events to travel from the source to the drain such that there is no velocity saturation is that experimental data show that they are not totally contradictory. The author's experience is that the saturation velocity v_{sat} in Eqs. (3.7) and (3.8) can be replaced by an effective saturation velocity v_{sat_eff} as follows.

$$I_{ds} = (v_{sat\_eff} W C_{ox})(V_{GS} - V_{th,sat}) \tag{3.9}$$

$$g_m = v_{sat\_eff} W C_{ox} \tag{3.10}$$

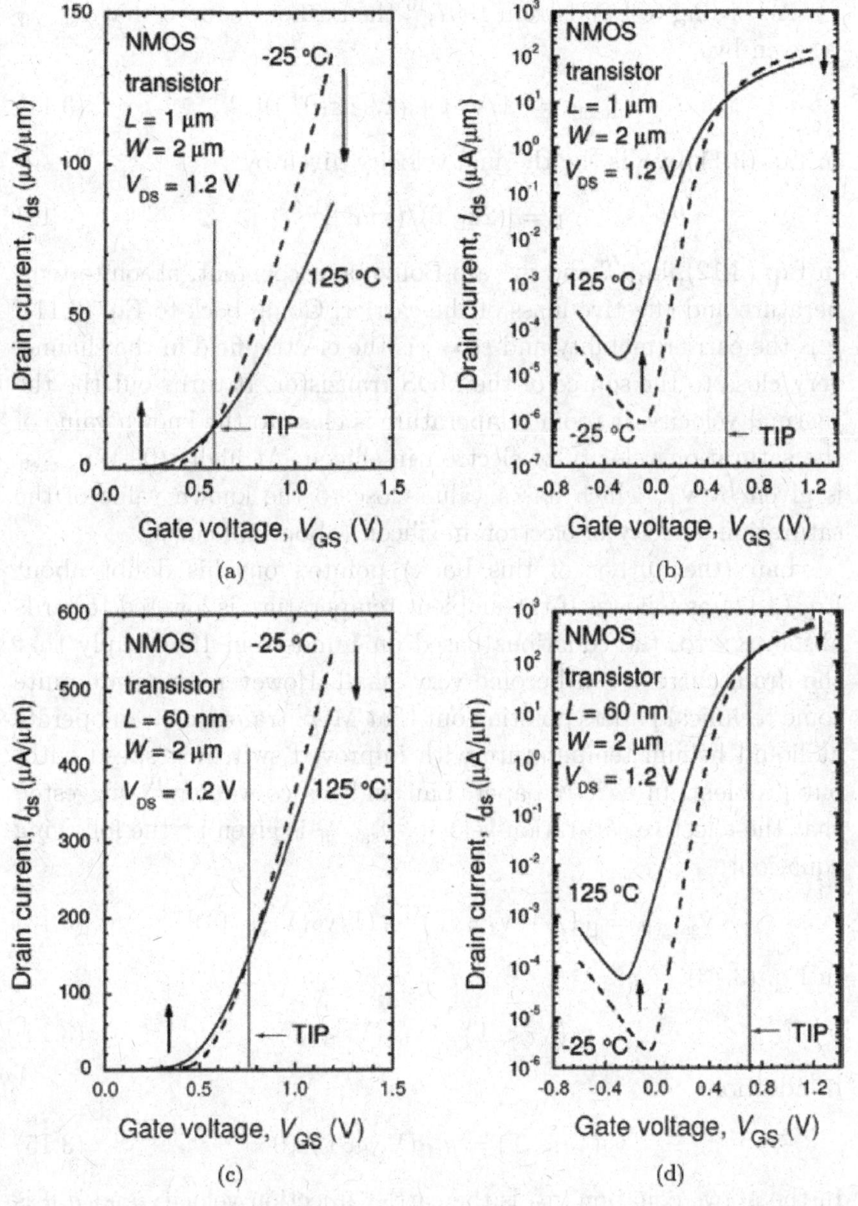

Fig. 3.7 Experimental data show that the drain current is usually smaller at higher temperature according to Yang *et al.*[15] Note: TIP stands temperature independent point. (P. Yang was a PhD student of the author and so this figure is part of the work of the author.)

According to Lundstrom 1997,[16] the author deduced that v_{sat_eff} is given by

$$v_{sat\_eff} = [(1/v_T) + (1/(\mu\varepsilon(0^+))]^{-1} \tag{3.11}$$

In Eq. (3.11), v_T is the thermal velocity given by

$$v_T = [(2k_BT)/(\pi m^*)]^{1/2} \tag{3.12}$$

In Eq. (3.12), k_B, T and m^* are Boltzmann constant, absolute temperature and effective mass of the carrier. Going back to Eq. (3.11), μ is the carrier mobility and $\varepsilon(0+)$ is the electric field in the channel very close to the source of the MOS transistor. It turns out the the thermal velocity at room temperature is close to the known value of the saturation velocity of electron in silicon. At high $\varepsilon(0+)$, v_{sat_eff} is given by v_T, which has a value close to the known value of the saturation velocity of electron in silicon (about 10^7 cm/s).

Lau (the author of this book) pointed out his doubt about Eq. (3.11) as follows. If the ambient temperature is lowered towards absolute zero, the equations based on Lundstrom 1997 imply that the drain current will become very small. However, there are quite some technical papers pointing out that MOS transistors can operate at liquid helium temperature with improved switching speed without problem. In a 2008 paper, Lau and his co-workers[17] suggested that the effective saturation velocity v_{sat_eff} is given by the following equation:

$$v_{sat\_eff} = [(1/v_1(V_{GS}, T)) + (1/v_2(V_{GS}, T))]^{-1} \tag{3.13}$$

In Eq. (3.13),

$$v_1(V_{GS}, T) = v_{inj}(V_{GS}, T) \tag{3.14}$$

In addition,

$$v_2(V_{GS}, T) = \mu_{eff}(V_{GS}, T)\varepsilon(0^+) \tag{3.15}$$

In the above equation v_{inj} is the carrier injection velocity and μ_{eff} is the effective carrier mobility. Lau *et al.* considered both quantities to be functions of both the gate to source voltage V_{GS} and the absolute temperature.

In a 2008 paper, Lau and his co-workers suggested that there is some sort of "hot carrier effect" existing in an MOS transistor such that v_T in Eq. (3.11) should be replaced by $v_1(V_{GS}, T) = v_{inj}(V_{GS}, T)$ and v_1 can be larger than the thermal velocity v_T because of some sort of "hot carrier effect". That is the carriers (electrons or holes) may be effectively "hotter" than the ambient temperature. The author believed that there is some sort of "hot carrier effect" because of the gate voltage and drain voltage applied to an MOS transistor. As shown in Fig. 3.8, the author believes that $v_1(V_{GS}, T)$ is more or less constant and equal to the thermal velocity at the absolute temperature T; when V_{GS} is increase, $v_1(V_{GS},T)$ can increase significantly. As shown in Fig. 3.8, the author believes that $v_2(V_{GS}, T)$ first increases with the increase of V_{GS} and then slightly decreases with the further increase of V_{GS}. The overall effect is that the effective saturation velocity first increases with the increase of V_{GS} and then saturates with the further increase of V_{GS}. Since $v_1(V_{GS}, T)$ can be significantly larger than the thermal velocity, the decrease of the ambient temperature to liquid helium temperature may not cause a decrease in the drain current. In fact, at liquid helium temperature, the mobility becomes better such that $v_2(V_{GS}, T)$ can be

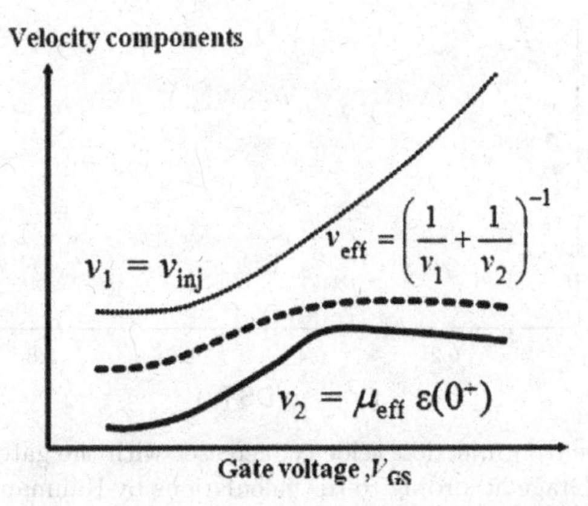

Fig. 3.8 A schematic diagram showing the relationship of $v_1(V_{GS}, T)$, $v_2(V_{GS}, T)$ and the effective saturation velocity as a function of V_{GS}. This figure is similar to Fig. 13 according to Lau *et al.* 2008.[17] (This figure comes from the work of the author.)

larger, resulting in a larger effective saturation velocity compared to operation at room temperature.

As discussed above, Lau and his co-workers suggested that there is some sort of "hot carrier effect" existing in an MOS transistor such that the carrier injection velocity can be larger than the thermal velocity. In 1997, Lundstrom treated the injection velocity as a quantity equal to the thermal velocity. Subsequently, Lundstrom and his co-workers appeared to have slightly changed their mind. v_T is replaced by v_{inj} (injection velocity), which is equal to v_T at low gate voltage but can be greater than v_T at high gate voltage. As shown in Fig. 3.9, calculations by Rahman, Guo, Datta and Lundstrom[18] showed that the electron injection velocity is not a constant but it can increase with the increase of the gate voltage and also with the increase of the drain voltage.

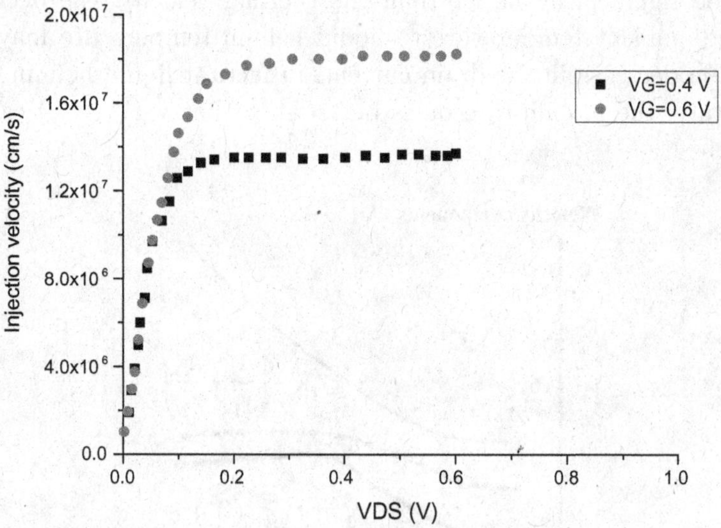

Fig. 3.9 The electron injection velocity increases with the gate voltage and also with the drain voltage according to the calculations by Rahman, Guo, Datta and Lundstrom. This figure is similar to Fig. 5(b) according to Rahman, Guo, Datta and Lundstrom 2003.[18] (Modified from Fig. 5(b) in A. Rahman, J. Guo, S. Datta and M.S. Lundstrom, "Theory of ballistic nanotransistors", *IEEE Trans. Electron Dev.*, vol. 50, no. 9 (September 2003), pp. 1853–1864.)

As discussed above, in 1994, Natori proposed a pure ballistric transport theory. Subsequently, Natori and his co-workers appeared to have been converted to the quasi-ballistic transport theory, as seen in their papers after 2000, for example, Natori 2008.[19] Natori and his co-workers also believed that v_{inj} is equal to v_T at low gate voltage but can be greater than v_T at high gate voltage and explained this by some sort of "charge degeneracy".[20] Note: The validity of equations (3.9) and (3.10) comes from the author's observation that the drain current versus gate voltage characteristics is quite linear for modern state-of-the-art MOS transistors with very small gate lengths, for example, 0.12 μm. Equations (3.5) and (3.6) are still valid for modern state-of-the-art MOS transistors with long gate lengths, for example, 10 μm. Of course, Eqs. (3.5) and (3.6) are valid for MOS transistors fabricated by old CMOS technology. However, the author observes that Eqs. (3.5) and (3.6) are still quite useful for modern state-of-the-art thin film transistors (TFTs); the square root of the drain current versus gate voltage characteristics is quite linear for modern state-of-the-art thin film transistors (TFTs).

Experimentally, it has been observed that the effective saturation velocity can be smaller or larger than the expected value of the saturation velocity. Up to now, the value of the effective saturation velocity found by experiment is still close to the expected value of the saturation velocity according to Yang *et al.*[21] However, for engineers, the concept of the effective saturation velocity allows the improvement of the effective saturation velocity by an improvement of the mobility. This is because the effective saturation velocity can be a function of the mobility and in general the effective saturation velocity increases with the increase of the mobility. As shown in Fig. 3.10, the effective saturation velocity increases with the increase of low field mobility for both electrons and holes in silicon according to data from Ohba and Mizuno.[22] Tatsumura *et al.* also pointed out that the effective saturation velocity increases with the increase of mobility in 2009.[23] For the old theory based on velocity saturation, the carrier drift velocity in MOS transistors will not increase with the decrease of channel length once "velocity saturation" is reached. However, as

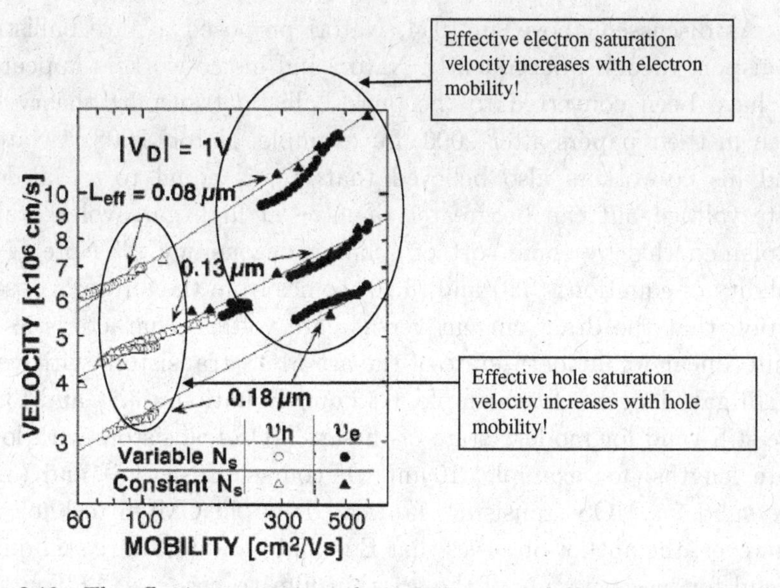

Fig. 3.10 The effective saturation velocity increases with the increase of low field mobility for both electrons and holes in silicon according to Ohba and Mizuno.[22] (Modified from Fig. 7 in R. Ohba and T. Mizuno, "Nonstationary electron/hole transport in sub-0.1 μm MOS devices: correlation with mobility and low-power CMOS applications", *IEEE Trans. Electron Dev.*, vol. 48, no. 2 (Feb. 2001), pp. 338–343.)

shown in Fig. 3.11, the effective saturation velocity increases with the decrease of channel length for both electrons and holes in silicon according to data from Ohba and Mizuno.[22] For engineers, theories like ballistic transport or quasi-ballistic transport may not be really useful. For engineers, the key point for higher switching speed is higher drive current which can be achieved by "higher mobility" and "shorter channel length"!

From another point of view, when the saturation velocity is replaced by the effective saturation velocity, the value of the effective saturation velocity can be used as a fitting parameter for actual experimental data fitting. This approach may be useful for circuit simulation. Up to now, old models based on low field mobility and high field saturation velocity (used as fitting parameter) still work quite well. Simulation engineers may not have to care whether the simulation model used is old-fashioned or fashionable. For example,

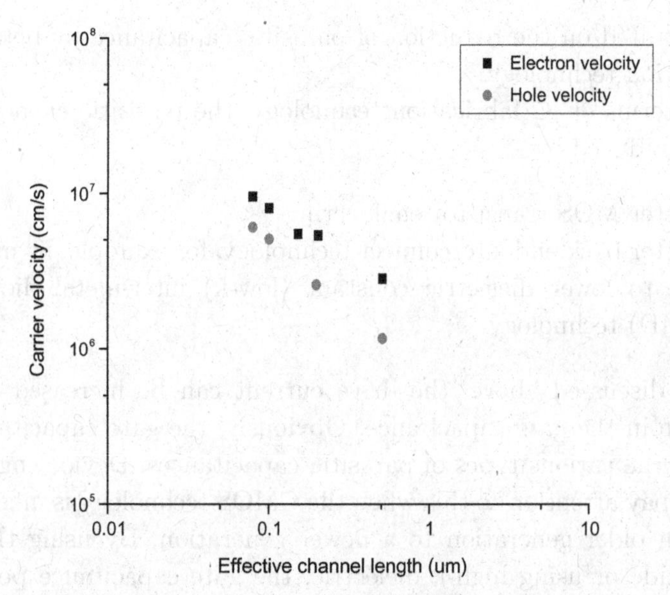

Fig. 3.11 The effective saturation velocity increases with the decrease of channel length for both electrons and holes in silicon according to Ohba and Mizuno.[22] The absolute value of the drain voltage was 1 V. The absolute value of the difference of the gate voltage and the threshold voltage was 1 V. (Modified from Fig. 5 in R. Ohba and T. Mizuno, "Nonstationary electron/hole transport in sub-0.1 μm MOS devices: correlation with mobility and low-power CMOS applications", *IEEE Trans. Electron Dev.*, vol. 48, no. 2 (Feb. 2001), pp. 338–343.)

there was an old-fashioned model proposed by Hauser in 2005 without using any ballistic or quasi-ballistic carrier transport theory.[24] In fact, Jeong *et al.*[25] pointed out: "Finally, the scattering model helps explain why conventional MOSFET models based on drift-diffusion concepts continue to work surprisingly well for nanoscale channel lengths."

3.3.2 *Decrease parasitic capacitance*

There is parasitic capacitance associated with the front-end and also parasitic capacitance associated with the back-end. Both types of parasitic capacitance can be reduced by (1) better IC circuit design and (2) better IC fabrication technology. The focus of this book is actually IC fabrication technology and so our discussion will be concentrated on the reduction of parasitic capacitance by better IC fabrication technology.

In terms of IC fabrication technology, the parasitic capacitance can be reduced by

(1) Better MOS transistor engineering
(2) Better back-end interconnect technology, for example, by migrating to lower dielectric constant (low-k) inter-metal dielectric (IMD) technology

As discussed above, the drive current can be increased by an increase in the gate capacitance. Obviously, the gate capacitance is part of the various types of parasitic capacitances. Device engineers should pay attention to this when the CMOS technology is migrating from an older generation to a newer generation. By using thinner gate oxide or using high-k dielectric, the gate capacitance per unit area is increased, resulting in an increase in the drive current divided by the MOS transistor width; to prevent a degradation in switching speed due to a bigger parasitic capacitance, proper scaling down of the channel length can decrease the absolute gate capacitance back to an acceptable level such that the overall effect of migrating to a new generation of CMOS technology can be an improvement in the switching speed.

3.4 CMOS Technology Improvement by Mobility Improvement of Silicon

In vacuum, electrons move according to Newton's second law of motion. (This is the case when the electron velocity is significantly smaller than the velocity of light. When the electron velocity is close to the velocity of light, the theory of relativity is required instead of Newton's second law of motion.) Under a constant electric field, electrons will not move at a constant velocity; instead, electrons will accelerate in a non-stop manner as dictated by Newton's second law of motion. This is known as "ballistic" transport. In a semiconductor, electrons can be scattered by phonons or impurities, etc. Under a constant electric field, when there is scattering, electrons will move at a constant drift velocity after statistical averaging. When the electric field is low, the drift velocity is proportional to the electric field and the proportionality constant is known as "mobility".

For an MOS transistor, there are 4 main mechanisms of scattering:

(1) Coulombic scattering
(2) Phonon scattering
(3) Interface roughness scattering
(4) Remote charge scattering

Takagi *et al.*[26,27] made a detailed study on the first three mechanism of carrier scattering. Figure 3.12 shows the first three

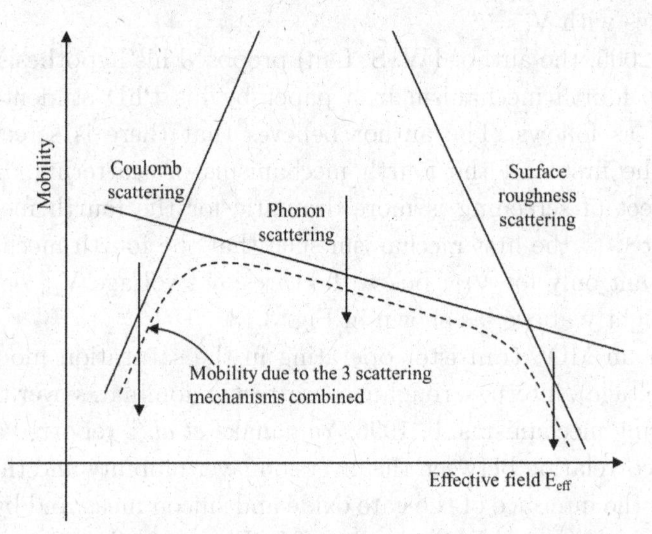

Fig. 3.12 Three different mechanisms of scattering, which can affect the low-field mobility in the channel region of an MOS transistor, as a function of the effective field E_{eff}, which increases with the increase of the gate-to-source voltage V_{GS}. The 3 arrows show that the mobility for the 3 scattering mechanisms decrease with temperature such that the mobility due to the 3 scattering mechanisms combined decrease with the increase in temperature. (Modified significantly from Fig. 3 in S. Takagi, A. Toriumi, M. Iwase and H. Tango, "On the universality of inversion layer mobility in Si MOSFET's: Part I — Effects of substrate impurity concentration", *IEEE Trans. Electron Dev.*, vol. 41, no. 12 (Dec. 1994), pp. 2357–2362.)

mechanisms of scattering in an MOS transistor as a function of the effective field E_{eff}, which increases with the increase of the gate-to-source voltage V_{GS}.

$$E_{eff} = (q/\varepsilon_{Si})(N_{dpl} + \eta N_s) \quad (3.16)$$

In Eq. (3.16), q is the electronic charge, ε_{Si} is the permittivity of silicon, N_{dpl} is the depleted net dopant concentration per unit area and N_s is the inversion carrier concentration per unit area. (Note: The net dopant concentration is the difference between the concentration of donors and that of acceptors.) In addition, η is an important parameter; it is taken to be 1/2 when electron mobility is to be studied while it is taken to be 1/3 when hole mobility is to be studied. Since $N_{dpl} + \eta N_s$ increases with the gate-to-source voltage V_{GS}, E_{eff} increases with V_{GS}.

In 2005, the author (W. S. Lau) proposed his hypothesis regarding the fourth mechanism in a paper by his PhD student (C. W. Eng)[28] as follows. The author believes that there is screening for both the first and the fourth mechanisms of scattering. However, the effect of screening is more dramatic for the fourth mechanism compared to the first mechanism such that the fourth mechanism is important only for V_{GS} below the threshold voltage V_{th} or at least only slightly above, as shown in Fig. 3.13.

For an MOS transistor operating in the saturation mode (large gate voltage), interface roughness scattering dominates over the other scattering mechanisms. In 1996, Yamanake *et al.*[29] reported an interesting correlation between the inversion layer mobility and the roughness at the interface of the gate oxide and silicon measured by atomic force microscope (AFM). As shown in Fig. 3.14, the inversion layer mobility was shown to decrease with the measured RMS roughness. (Note: RMS is root mean square.)

3.4.1 *Faster CMOS by operation at lower temperature*

The carrier mobility can be improved by various methods. For example, carrier mobility can be improved by operating CMOS integrated circuits at low temperatures. For Coulombic scattering, scattering is

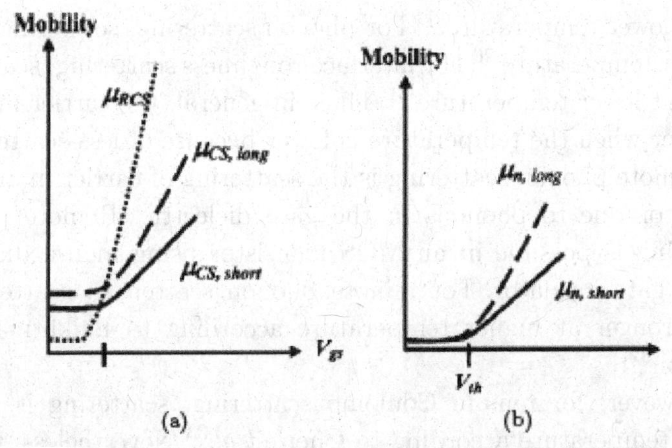

Fig. 3.13 (a) Remote charge scattering (RCS) and Coulombic scattering components of mobility close to threshold. (b) Mobility of short and long MOS transistors close to threshold according to Lau's hypothesis.[28] (C.W. Eng was a PhD student of the author and so this figure is part of the work of the author.)

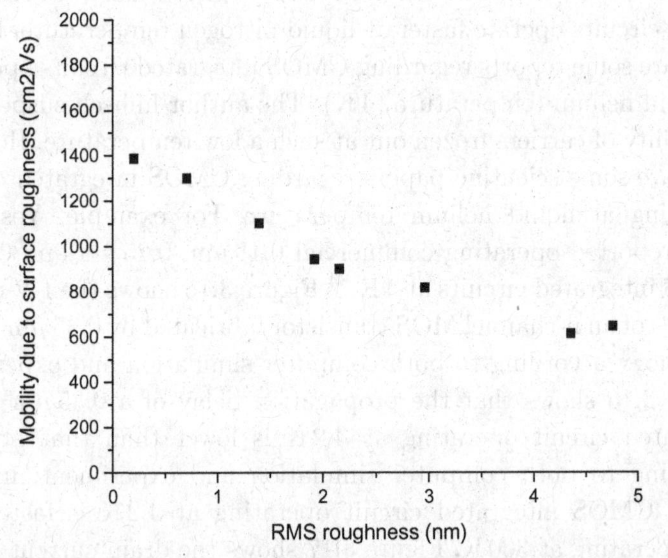

Fig. 3.14 The inversion layer mobility was shown to decrease with the measured RMS roughness by using an atomic force microscope (AFM). (Modified from Fig. 2 in T. Yamanaka, S. J. Fang, H.-C. Lin, J. P. Snyder and C. R. Helms, "Correlation between inversion layer mobility and surface roughness measured by AFM", *IEEE Electron Dev. Lett.*, vol. 17, no. 4 (April 1996), pp. 178–180.)

less at lower temperature.[30] For phonon scattering, scattering is less at lower temperature.[26] For interface roughness scattering, scattering is less at lower temperature.[31] Thus, in general, the carrier mobility is higher when the temperature is lower because of less scattering.

Remote phonon scattering is the scattering of carrier in an MOS transistor due to phonons in the gate dielectric. Remote phonon scattering is possible in an MOS transistor using high-k dielectric as the gate insulator. For remote phonon scattering, scattering is also stronger at higher temperature according to Laikhtman and Solomon[32]!

However, for remote Coulomb scattering, scattering is less at higher temperature according to Chen *et al.*[33] Nevertheless, remote Coulomb scattering is only important at weak inversion (gate voltage below threshold voltage) and so it can be said that in general the carrier mobility is higher when the temperature is reduced.

There are some reports regarding CMOS integrated circuits operating at liquid nitrogen temperature (77 K); usually, CMOS integrated circuits operate faster at liquid nitrogen temperature. In fact, there are some reports regarding CMOS integrated circuits operating at liquid helium temperature (4 K). The author himself suspects the possibility of carriers frozen out at such a low temperature. However, there are some scientific papers regarding CMOS integrated circuits operating at liquid helium temperature. For example, Yoshikawa *et al.* reported operating commercial 0.18 μm, 0.25 μm and 0.35 μm CMOS integrated circuits at 4 K.[34] Figure 3.15 shows the I-V characteristics of an n-channel MOS transistor fabricated by 0.35 μm CMOS technology according to both computer simulation and experiment. Figure 3.16 shows that the propagation delay of a 0.35 μm CMOS integrated circuit operating at 4.2 K is lower than that at 300 K according to both computer simulation and experiment; in other words, CMOS integrated circuit operating at 4.2 K is faster than that operating at 300 K. Figure 3.17 shows the drain current I_D versus gate-to-source voltage V_{GS} characteristics of an n-channel MOS transistor fabricated by 90 nm CMOS technology at 4.2 K and 300 K according to Hong *et al.*[35] The readers should be able to notice that there is a cross-over of the two I_D-V_{GS} characteristics. This cross-over

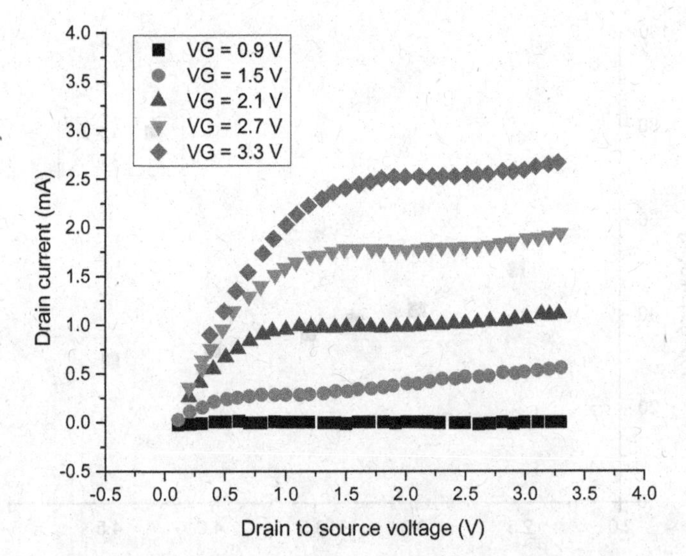

Fig. 3.15 Measured I-V characteristics of a 0.35 μm NMOS device at 4.2 K according to Yoshikawa *et al.*[34] Gate width was 4.45 μm. (Modified from Fig. 1 in N. Yoshikawa, T. Tomida, M. Tokuda, Q. Liu, X. Meng, S. R. Whiteley and T. Van Duzer, "Characterization of 4 K CMOS devices and circuits for hybrid Josephson-CMOS systems", *IEEE Trans. Appl. Superconductivity*, vol. 15, no. 2 (June 2005), pp. 267–271.)

can be observed in general for two I_D-V_{GS} characteristics at two different temperatures. This is due to the lower I_D at high V_{GS} (device turned on) because of the lower mobility at high temperature and the higher I_D at low V_{GS} (device turned off) because of the degraded subthreshold swing at high temperature. Thus operation at low temperature improves both the on current and the off current of the MOS transistors.

In addition, the cross-over of two I_D-V_{GS} characteristics at two different temperatures creates a temperature independent point (TIP) as shown in Fig. 3.7. If this TIP is exploited properly, there is some potential to reduce the sensitivity of MOS devices to changing temperature as discussed by Yang *et al.* 2010.[15] (Note: Peizhen Yang was a PhD student of the author.)

As discussed above, lower temperature leads to faster CMOS operation. Thus any IC packaging technology which can improve

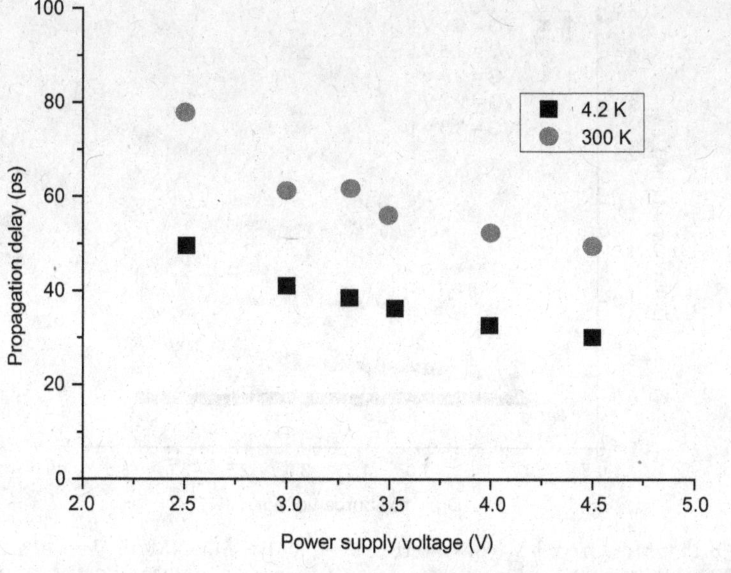

Fig. 3.16 Measurement results of inverter propagation delay fabricated by a 0.35 μm CMOS technology with nominal L = 0.35 μm and W = 4.45 μm operating at 4.2 K and 300 K according to Yoshikawa *et al.*[34] It can be easily seen that operation at 4.2 K is much faster than operation at 300 K. (Modified from Fig. 3 in N. Yoshikawa, T. Tomida, M. Tokuda, Q. Liu, X. Meng, S. R. Whiteley and T. Van Duzer, "Characterization of 4 K CMOS devices and circuits for hybrid Josephson-CMOS systems", *IEEE Trans. Appl. Superconductivity*, vol. 15, no. 2 (June 2005), pp. 267–271.)

heat removal will slightly improve switching speed. Nanotechnology has attracted world-wide attention for science and technology. For example, carbon nanotubes are one of the many nano materials which have attracted world-wide attention. The discovery of carbon nanotubes is usually attributed to Sumio Iijima (1939–) in 1991. Recently, carbon nanotube (CNT) technology has been investigated for device applications. In fact, CNT transistor has the potential to perform better than Si-based transistors according to Iijima *et al.* in 2007.[36] However, the author feels that CNT device technology is very much not mature enough for IC manufacturing but CNT technology has also been investigated as a novel approach to improve heat removal for IC packaging technology and the author feels that

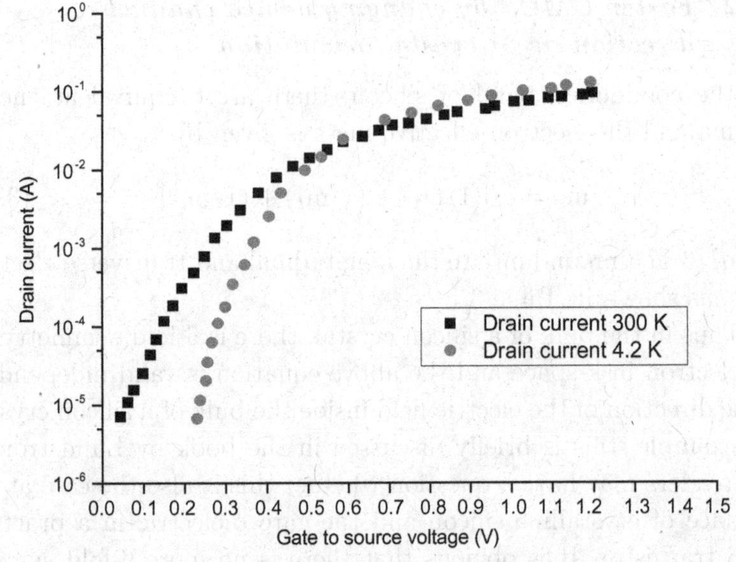

Fig. 3.17 Experimental drain current versus gate-to-source voltage of an n-channel MOS transistor fabricated by 90 nm CMOS technology operating at 4.2 K and 300 K according to Hong *et al.*[35] (Modified from Fig. 1 in S.-H. Hong, G.-B. Choi, R.-H. Baek, H.-S. Kang, S.-W. Jung and Y.-H. Jeong "Low-temperature performance of nanoscale MOSFET for deep-space RF applications", *IEEE Electron Dev. Lett.*, vol. 29, no. 7 (July 2008), pp. 775–777.)

CNT technology for better heat removal may be much more compatible with IC manufacturing. In 1995, Ruoff and Lorents speculated that CNT may have thermal conductivity as high as diamond or graphite.[37] Subsequently in 1999, Hone *et al.* actually measured the thermal conductivity of CNT and claimed that CNT may really have thermal conductivity as high as diamond or graphite.[38] In addition, Berber *et al.* pointed out the unusually high thermal conductivity for carbon nanotubes in 2000.[39] Subsequently, there are more reports regarding the exploitation of the high thermal conductivity of carbon nanotubes for thermal management, for example, in IC technology.[40–42] Thus the application of CNT heat removal technology for IC packaging is more likely to make CMOS integrated circuits operating faster compared to the application of CNT device technology.

3.4.2 *Faster CMOS by changing device channel direction or Si crystal orientation*

For the conduction band of silicon, there are 6 equivalent energy minima and the electron effective mass is given by

$$m_e = 3/[(1/m_l) + (1/m_t) + (1/m_t)] \tag{3.17}$$

In Eq. (3.17), m_l and m_t are the longitudinal and transverse electron mass, as shown in Table 3.1.

Thus in the bulk of a silicon crystal, there is 6-fold symmetry for the electron in k-space and the above equation is valid independent of the direction of the electric field inside the bulk of a silicon crystal. For example, this is briefly discussed in the book by Lundstrom.[43] The readers may have a question whether this is also the case at the interface of crystalline silicon and the gate dielectric in a practical MOS transistor. It is obvious that there is no more 6-fold symmetry and the electron effective mass may depend on the silicon wafer orientation and the choice of the channel direction.

As shown in Fig. 3.18, in 2004, Chang *et al.* (IBM)[44] pointed out that the best electron mobility for n-channel MOS transistors was experimentally observed for (100) silicon wafers compared to (111) or (110) silicon wafers while the best hole mobility for p-channel MOS transistors was experimentally observed for (110) silicon wafers compared to (100) or (111) silicon wafers.

As shown in Fig. 3.19, in 2006, Yang *et al.* (IBM)[45] pointed out that the best electron mobility for n-channel MOS transistors was experimentally observed for (100) silicon wafers compared to (111) or (110) silicon wafers. In 2010, the author and co-workers showed that at least for (100) oriented silicon wafer, the performance of n-channel MOS transistors is independent of the MOS channel

Table 3.1. Longitudinal and transverse electron effective mass in silicon in units of the free electron mass.

Longitudinal electron effective mass m_l	Transverse electron effective mass m_t
0.91	0.19

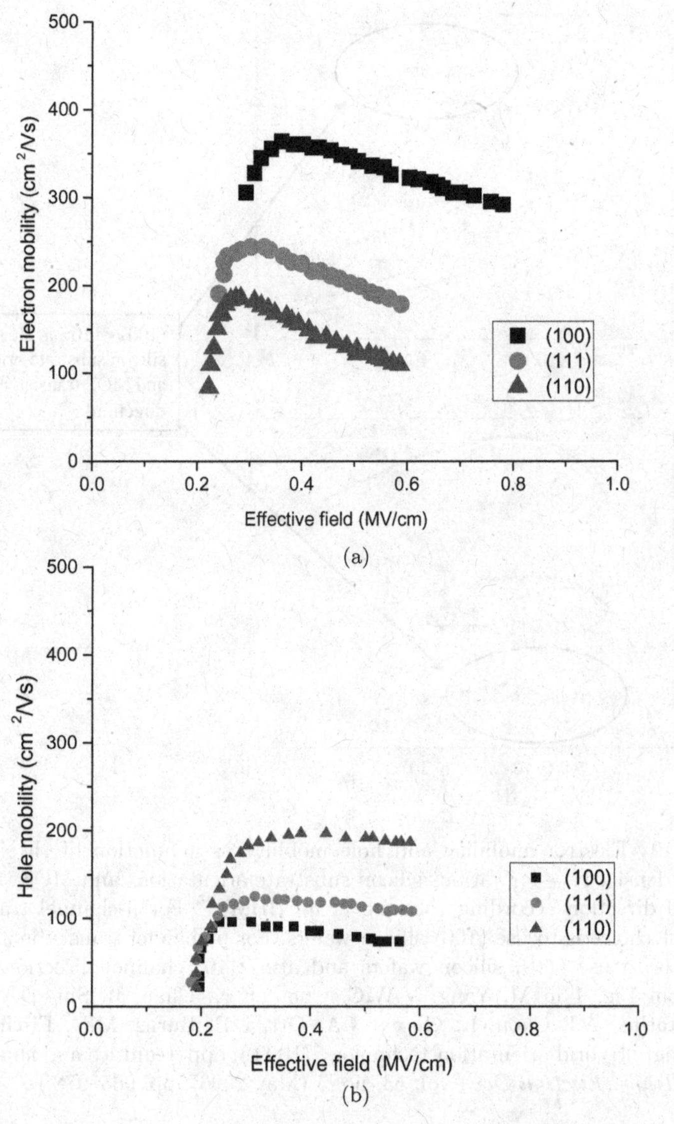

Fig. 3.18 (a) Electron and (b) hole mobility for (100), (110) and (111) silicon surface orientations as a function of the effective field according to Chang *et al.* (IBM).[44] Note: The direction of current flow was assumed to be perpendicular to the wafer notch. (Modified from Fig. 1 in L. Chang, M. Ieong and M. Yang, "CMOS circuit performance enhancement by surface orientation optimization", *IEEE Trans. Electron Dev.*, vol. 51, no. 10 (Oct. 2004), pp. 1621–1627.)

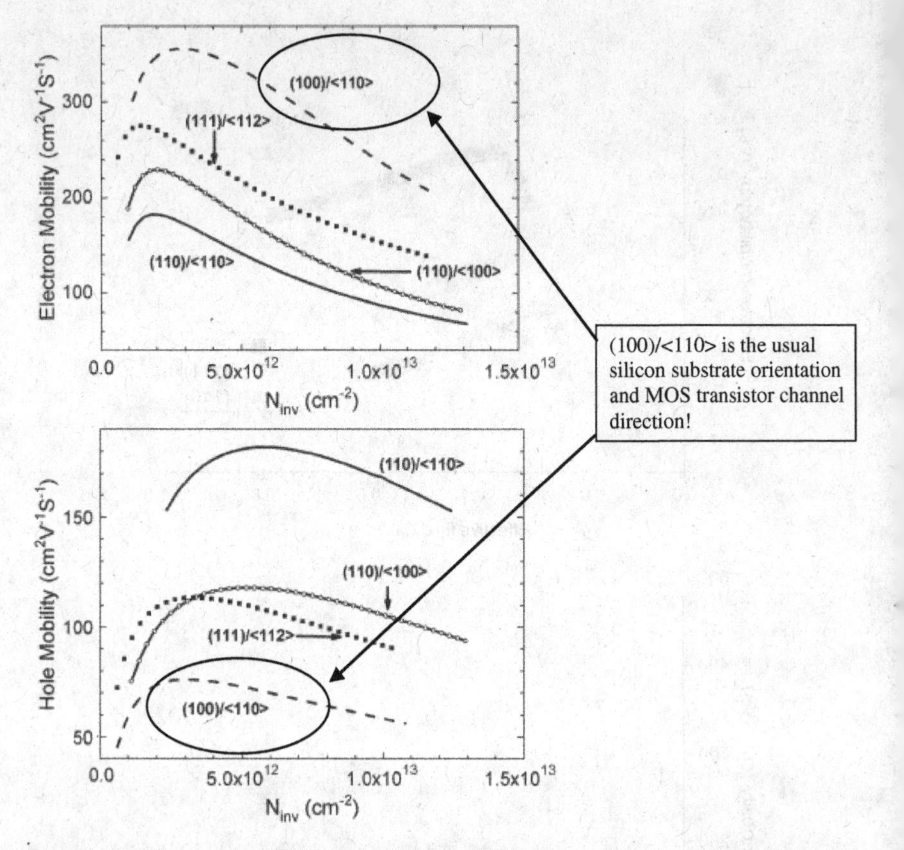

Fig. 3.19 Electron mobility and hole mobility as a function of the inversion carrier density N_{inv} for various silicon substrate orientations and MOS transistor channel direction according to Yang *et al.* (IBM).[45] For n-channel transistors, the best choice is to use (100) silicon wafers. For p-channel transistors, the best choice is to use (110) silicon wafers and also ⟨110⟩ channel direction. (Modified from Fig. 1 in M. Yang, V.W.C. Chan, K.K. Chan, L. Shi, D.M. Fried, J.H. Stathis, A.I. Chou, E. Gusev, J.A. Ott, L.E. Burns, M.V. Fischetti and M. Ieong, "Hybrid orientation technology (HOT): opportunities and challenges", *IEEE Trans. Electron Dev.*, vol. 53, no. 5 (May 2006), pp. 965–978.)

direction because the electron effective mass is independent of the MOS channel direction.[46]

For example, the electron mobility is the same for the traditional ⟨110⟩ channel orientation or the ⟨100⟩ channel orientation for n-channel MOS transistors fabricated on (100) oriented silicon wafers. However, the hole mobility can be significantly improved by switching from the traditional ⟨110⟩ orientation to the ⟨100⟩ orientation for p-channel MOS transistors fabricated on (100) oriented silicon wafers. The physics behind the above experimental observations is the band structure of silicon. As shown in Fig. 3.19, Yang *et al.* (IBM)[45] pointed out that the best hole mobility for p-channel MOS transistors was experimentally observed for (110) silicon wafers with ⟨110⟩ channel direction. Since the best electron mobility occurs for (100) oriented silicon wafer but the best hole mobility occurs for (110) oriented silicon wafer. One approach is to use the hybrid orientation technology (HOT) according to Yang *et al.* (IBM), as shown in Fig. 3.20.[45]

Another approach is to use (100) silicon wafers for the best electron mobility and to vary the p-channel MOS transistor channel

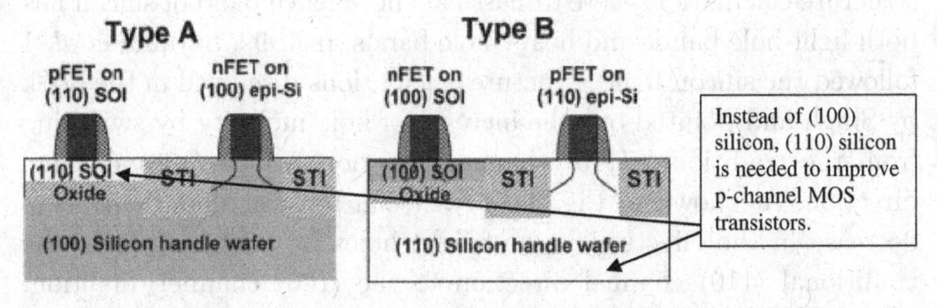

Fig. 3.20 Schematic cross sections of n-channel and p-channel MOS transistors fabricated by CMOS technology using hybrid orientation technology (HOT). There are two types according to Yang *et al.* (IBM):[45] type A with p-channel transistors on (110) SOI and n-channel transistors on (100) silicon epitaxial layer, and type B with n-channel transistors on (100) SOI and p-channel transistors on (110) silicon epitaxial layer. Best electron and hole mobilities can be achieved. (Modified from Fig. 2 in M. Yang, V.W.C. Chan, K.K. Chan, L. Shi, D.M. Fried, J.H. Stathis, A.I. Chou, E. Gusev, J.A. Ott, L.E. Burns, M.V. Fischetti and M. Ieong, "Hybrid orientation technology (HOT): opportunities and challenges", *IEEE Trans. Electron Dev.*, vol. 53, no. 5 (May 2006), pp. 965–978.)

direction for the best hole mobility achieved for (100) silicon wafers. The author will name this as the Cheaper Approach A. Another approach is to use (110) silicon wafers and the ⟨110⟩ channel direction for the best hole mobility and the best p-channel transistor performance but the electron mobility and so the n-channel transistor performance will be sacrificed. The author will name this as the Cheaper Approach B. These two approaches are obviously cheaper than the HOT approach proposed by Yang *et al.* (IBM).[45] In this book, the author will discuss more on the Cheaper Approach A.

The effect of silicon wafer surface orientation and the MOS transistor channel orientation on the electron and hole mobilities has been studied by various authors for several decades. Let us concentrate on the hole mobility in (100) silicon wafers. The increase of hole mobility by switching from the traditional ⟨110⟩ channel direction to the ⟨100⟩ channel direction can be understood only by a study of the band structure of silicon. In the book by Jasprit Singh,[47] silicon band structure calculations were discussed. The valence band of silicon has both light hole bands and heavy hole bands. In 1999, Sayama *et al.*[48] followed the silicon band structure calculations discussed in the book by Singh and pointed out the increase of hole mobility by switching from the traditional ⟨110⟩ channel direction to the ⟨100⟩ channel direction. As shown in Fig. 3.21(a), it can be seen that there is no decrease in the effective mass of light holes by switching from the traditional ⟨110⟩ channel direction to the ⟨100⟩ channel direction. However, as shown in Fig. 3.21(b), it can be seen that there is a decrease in the effective mass of heavy holes by switching from the traditional ⟨110⟩ channel direction to the ⟨100⟩ channel direction. Thus there is an overall effective decrease in the effective mass of holes by switching from the traditional ⟨110⟩ channel direction to the ⟨100⟩ channel direction, resulting in an increase in hole mobility. According to Singh's book, the light hole band becomes more isotropic at higher hole energy and the heavy hole band becomes more anisotropic at higher hole energy. The implication is that the effect of "channel orientation rotation" is more obvious at higher electric fields!

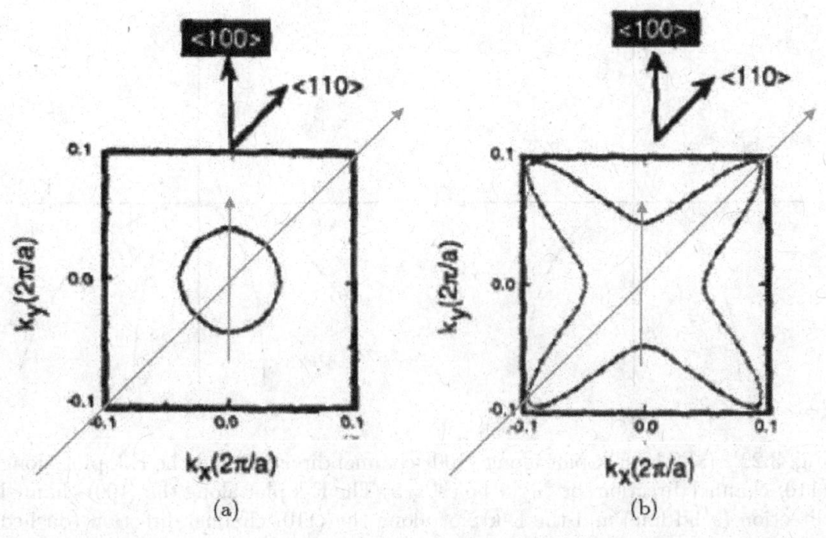

Fig. 3.21 A cross section of the constant energy surface at the $k_z = 0$ of (a) the light hole bands and (b) the heavy hole bands in silicon at 40 meV below the top of the valence band of silicon. (This figure has been created by a modification from Singh[47] and Sayama *et al.*[48]) (Modified from Fig. 1 and Fig. 2 in H. Sayama, Y.Nishida, H. Oda, T. Oishi, S. Shimizu, T. Kunikiyo, K. Sonoda, Y. Inoue and M. Inuishi, "Effect of ⟨100⟩ channel direction for high performance SCE immune pMOSFET with less than 0.15 μm gate length", *IEDM Technical Digest*, (1999), pp. 657–660.)

A better diagram than Fig. 3.21 can be found in the literature; for example, Fig. 17 in Skotnicki *et al.* 2008.[49] As seen in Fig. 17 in Skotnicki *et al.* 2008,[49] for p-channel transistors fabricated on (100) Si wafers, the hole effective mass is smaller for the ⟨100⟩ channel direction compared to the ⟨110⟩ channel direction, resulting in higher hole mobility for the ⟨100⟩ channel direction compared to the ⟨110⟩ channel direction. Thus, one simple way to improve hole mobility without degradation of the electron mobility is to rotate the (100) silicon wafer from the conventional ⟨110⟩ channel direction to the ⟨100⟩ channel direction.

Figure 3.22(a) shows the E-k plot along the ⟨100⟩ channel direction and the E-k plot along the ⟨110⟩ channel direction for "light holes"; there is no difference. Figure 3.22(b) shows the E-k plot along the ⟨100⟩ channel direction (solid line) and the E-k plot along the

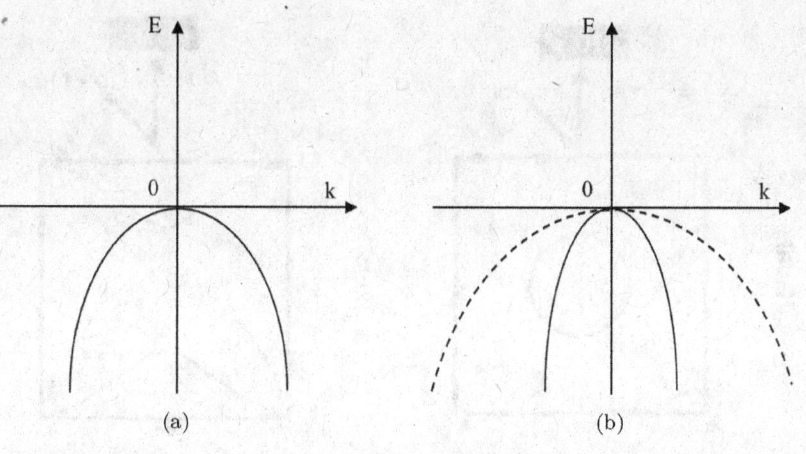

Fig. 3.22 (a) The E-k plot along ⟨100⟩ channel direction and the E-k plot along ⟨110⟩ channel direction for "light holes". (b) The E-k plot along the ⟨100⟩ channel direction (solid line) and the E-k plot along the ⟨110⟩ channel direction (dashed line) for "heavy holes". The readers should note that the ⟨100⟩ and ⟨110⟩ channel directions are shown as a shorter arrow and a longer arrow, respectively, in Fig. 3.21.

⟨110⟩ channel direction (dashed line) for "heavy holes"; the magnitude of the reciprocal of the second derivative of E with respect to k is much better for the ⟨100⟩ channel direction than the ⟨110⟩ channel direction.

Previously, the research group of the author has made a detailed study of the effect of switching from the traditional ⟨110⟩ channel direction to the ⟨100⟩ channel direction. The author pointed out that this switching also reduces the "lateral ion channeling" of boron, resulting in an increase in the effective channel length of p-channel MOS transistors.[50] A detailed study of the effect of switching from the traditional ⟨110⟩ channel direction to the ⟨100⟩ channel direction for both n-channel and p-channel MOS transistors on (100) silicon wafers was also published by the research group of the author[51]; there is also a reduction in the "lateral ion channeling" of arsenic but the effect is not as obvious as that of boron.

The readers should note why the ⟨110⟩ channel direction has been the traditional choice. This is related to the technology of cutting silicon wafers. The old way of cutting silicon wafers involved a "scribing" step by a sharp tool and then a "fracturing" step. For example, this was discussed by Misra and Finnie.[52] For silicon single crystal wafers, the best cleavage planes are the {111} family of planes. However, people wanted rectangular silicon chips; scribing along the {111} family of planes cannot produce rectangular silicon chips. For silicon single crystal wafers, the second and third best cleavage planes are the {110} family of planes and {100} family of planes according to Anner.[53] Scribing along the {110} family of planes can produce rectangular silicon chips. For CMOS technology, the process starts with a (100) silicon wafer. For silicon wafers with diameter smaller than 8 inches (200 mm), orientation flats are used to denote crystallographic orientations. For silicon wafers with diameter equal to or larger than 8 inches (200 mm), notches are used to denote crystallographic orientations. Traditionally, ⟨110⟩ notch is used; scribing along the {110} family of planes has been preferred. However, newer silicon wafer cutting technology like "diamond saw cutting" or "laser cutting" can cut through the silicon wafer without involving a "scribing" step by a sharp tool and then a "fracturing" step. Recently, silicon wafers with ⟨100⟩ notch are available. Thus the switching from the traditional ⟨110⟩ channel direction to the ⟨100⟩ channel direction simply involves the switching of starting wafers from (100) silicon wafers with the traditional ⟨110⟩ notch to (100) silicon wafers with the ⟨100⟩ notch; this will cause minimal change in the process technology involved.

3.4.3 *Faster CMOS by applying mechanical stress to silicon*

It is known that tensile stress can increase electron mobility in silicon while compressive stress can increase hole mobility in silicon. Mechanical stress is quite frequently applied to silicon by using a silicon nitride film deposited by plasma-enhanced chemical vapor deposition (PECVD). PECVD silicon nitride had been used as a contact etch stop layer (CESL) before "stress engineering of CMOS technology" became popular. Tungsten (W) contacts are used for the drain and source of MOS transistors in CMOS technology. The drain and source are usually covered by a layer of silicide. In order to save space, borderless drain/source contact technology was developed with W contacts right at the edge of the active silicon area adjacent to the shallow trench isolation (STI). For an ideal borderless W contact process, the W contact should not touch the STI. However, there is always some sort of misalignment during lithography such that the contact hole through the pre-metal dielectric (PMD) may have some overlap with the STI. The material inside STI is basically silicon dioxide. The PMD is usually some sort of oxide. The etch selectivity between the PMD and the STI oxide may be poor such that the contact hole etch may etch into the STI oxide. Subsequently, the W contact to the drain (or source) may cause a serious short-circuit from the drain (or source) to the well. For this kind of borderless drain/source contact technology, PECVD silicon nitride was used as a CESL. The etch selectivity between the PMD and the PECVD silicon nitride can be good enough such that the contact hole etch stops on top of the PECVD silicon nitride. Subsequently, the PECVD silicon nitride will be etched such that there is good electrical contact between the W and silicon (covered by silicide). This is the original purpose of the PECVD silicon nitride CESL. The application of a CESL for borderless drain or source contacts was discussed, for example, by Oishi *et al.*[54] Oishi *et al.* used a $SiO_2/Si_3N_4/SiO_2$ triple layer as CESL.[54] In general, some sort of PECVD silicon nitride or silicon oxynitride may be used.

Afterwards, it was recognized that the mechanical stress of this CESL can have a significant effect on the electrical characteristics of MOS transistors.[55] As shown in Fig. 3.23, the W contact to the drain or source touches both the active silicon area and the shallow trench isolation (STI) according to Liao *et al.*[56] (Note: For an ideal borderless contact technology, the W contact should touch only the active silicon area without touching the STI.) Liao *et al.* pointed out that the mild tensile stress in the CESL caused a slight improvement in the performance of n-channel transistors but also a slight degradation of the performance of p-channel transistors.[56]

By adjusting the deposition conditions, either tensile stress or compressive stress can be achieved for this CESL. The deposition

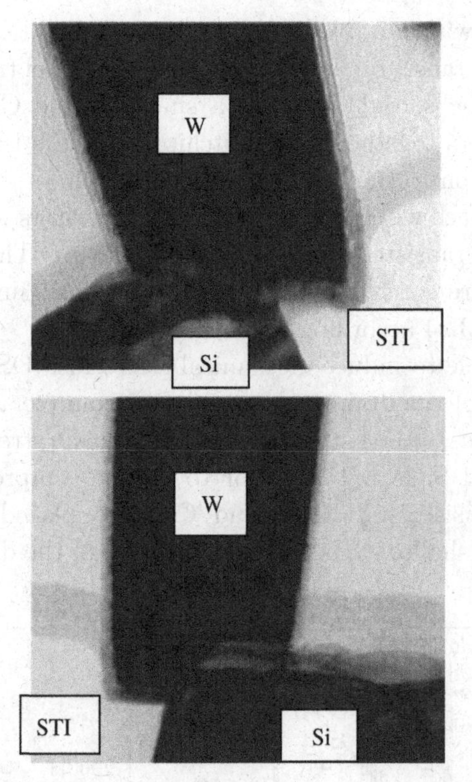

Fig. 3.23 XTEM (cross-sectional transmission electron microcopy) of wafers using (a) silicon nitride etch stop and (b) silicon oxynitride/silicon nitride etch stop according to Liao *et al.*[56] It can be seen that the W contact touched the silicon covered by a layer of silicide. In addition, the W contact also touched the shallow trench isolation (STI). (Modified from Fig. 2 in H. Liao, P. S. Lee, L. N. L. Goh, H. Liu, J. L. Sudijono, Q. Elgin and C. Sanford, "The impact of etch-stop layer for borderless contacts on deep submicron CMOS device performance — a comparative study", Thin Solid Films, vol. 462–463 (2004), pp. 29–33.)

conditions can be tuned to produce strong mechanical stress.[57] The dual stress liner (DSL) technology was to apply a tensile CESL to the n-channel MOS transistors and a compressive CESL to the p-channel MOS transistors to improve the electron mobility and the hole mobility of silicon respectively, resulting in a better CMOS technology.[58]

It is obvious that this DSL technology is an expensive technology. The cheaper approach is to deposit a tensile CESL on top of both n-channel and p-channel transistor. Using a tensile CESL alone on silicon wafers with traditional ⟨110⟩ channel direction will improve the n-channel transistors but degrade the p-channel transistors. This problem can be solved by using a single tensile CESL deposited on silicon wafers together with switching from the traditional ⟨110⟩ channel direction to the ⟨100⟩ channel direction.

This approach will improve n-channel transistors without degrading p-channel transistors according to Ishimaru.[59] The experimental results of the research group of the author show similar results as shown in Fig. 3.24 according to Yang *et al.*[51]

It is possible to induce mechanical stress in MOS transistors by modifications of the drain and source. For example, Ang *et al.* used SiC S/D (source/drain) stressor to induce tensile stress in n-channel transistors and SiGe S/D stressor to induce compressive stress in p-channel transistors.[60] Hsieh and Chung[61] placed the SiC S/D stressor in the shallow S/D extension instead of the deep S/D region

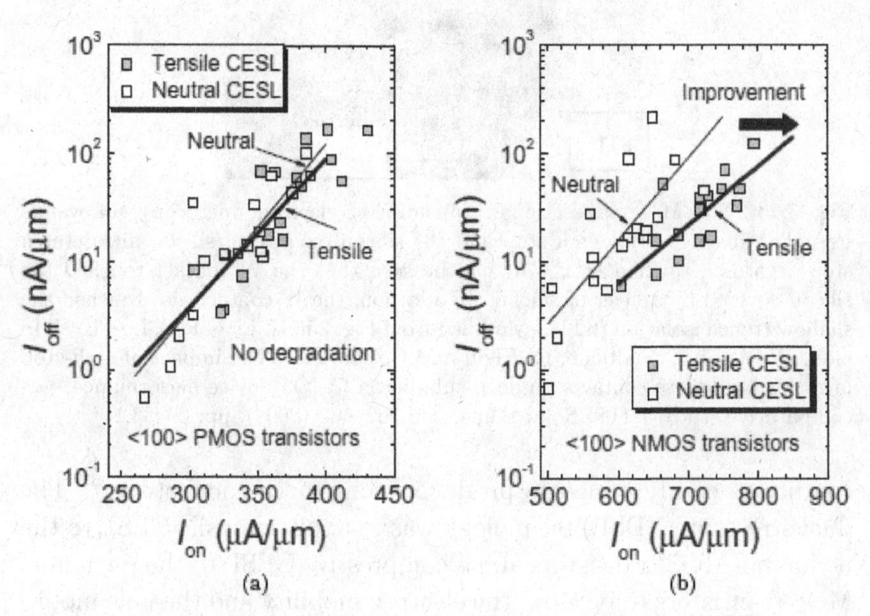

Fig. 3.24 A single tensile CESL deposited on silicon wafers together with switching from the traditional ⟨110⟩ channel direction to the ⟨100⟩ channel direction can improve n-channel transistors with no degradation for p-channel transistors according to Yang *et al.*[51] (P. Yang was a PhD student of the author and so this figure is part of the work of the author.)

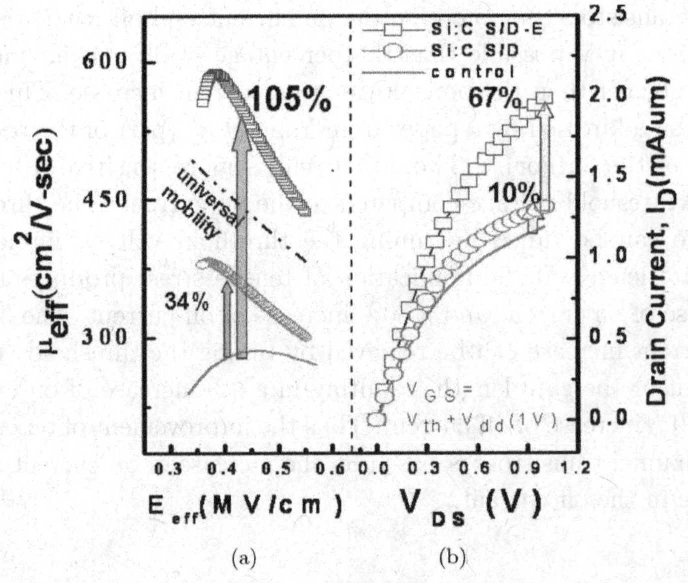

Fig. 3.25 (a) A plot of the effective mobility vs. the effective electric field shows an increase of 105% for Si:C S/D stressor in the shallow S/D extension region over the control device. For comparison, the same plot shows an increase of only 34% for conventional Si:C S/D stressor in the deep S/D region over the control device. (b) A plot of ID vs VGS (gate length = 40 nm) shows the enhancement of the drain current with over 67% increase in a device with Si:C S/D stressor in the shallow S/D extension over the control device. For comparison, the same plot shows an increase of only 10% in a device with conventional Si:C S/D stressor in the deep S/D region. Thus, a Si:C S/D stressor in the shallow S/D extension region is superior to a Si:C S/D stressor in the deep S/D region according to Hsieh and Chung.[61] (Reproduced with permission from E. R. Hsieh and S. S. Chung, "The proximity of the strain induced effect to improve the electron mobility in a silicon-carbon source-drain structure of n-channel metal-oxide-semiconductor field-effect transistors," *Appl. Phys. Lett.*, vol. 96, no. 9 (1 March 2010), pp. 093501-1 to 095501-3 with the permission of AIP Publishing. Copyright 2010 American Institute of Physics.)

of n-channel transistors, resulting in the SiC S/D stressor physically closer to the channel region and even better performance improvement, as shown in Fig. 3.25.

At this point, the author would like to point out that applying mechanical stress may change the on current and off current simultaneously. For example, the application of tensile stress to n-channel

MOS transistors may increase the on current and off current simultaneously; it is possible that the percentage of off current increase can be bigger than the percentage of on current increase. This issue has been addressed by a paper from Yang *et al.* (part of the research group of the author).[62] The off current is more sensitive to change in the threshold voltage compared to the on current. The threshold voltage can be tuned by tuning the threshold voltage implant or the gate length. If the application of tensile stress produces a 10% increase of on current and a 30% increase of off current. The 30% of off current increase can be removed by tuning the threshold voltage implant or the gate length, resulting in a 9% increase of on current and a 0% increase of off current. Thus the improvement of on current of n-channel transistors is based on the increase of on current for no change in the off current.

3.5 CMOS Technology Improvement by Reduction of Parasitic Capacitance

3.5.1 *Front-end techniques used to reduce parasitic capacitance*

3.5.1.1 Drain/source engineering

The capacitance of the deep n^+ drain/source to p-well and that of the deep p^+ drain/source to n-well is a source of parasitic capacitance which can degrade switching speed. For the deep n^+ drain/source to p-well junction, a more graded junction can be achieved by using an arsenic implant together with a phosphorus implant.[63] For the deep p^+ drain/source to n-well junction, a more graded junction can be achieved by using two boron implants.[64] Kim *et al.*[65] reduced the parasitic capacitance due to the deep n^+ drain/source to p-well and the deep p^+ drain/source to n-well by the techniques discussed above, resulting in a noticeable improvement in switching speed as shown in Fig. 3.26.

The application of carbon implant in the SDE/halo region has been discussed by various authors. (Note: SDE stands for source drain extension.) For example, it was discussed by Mineji and

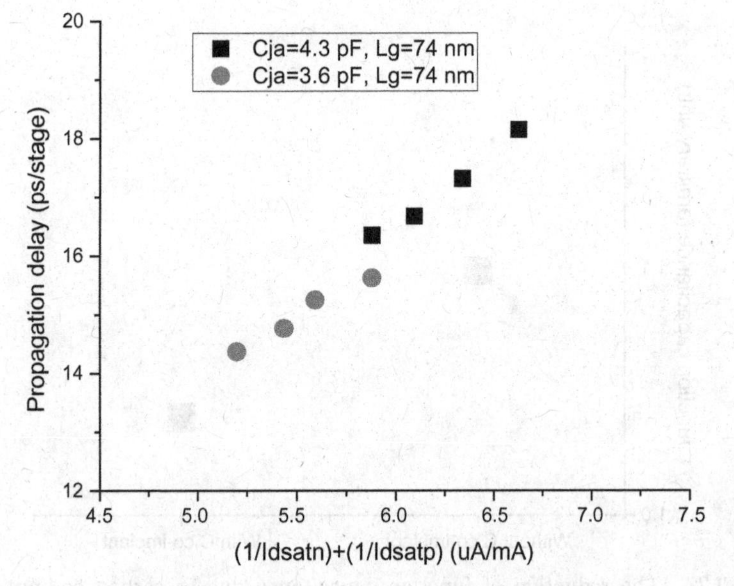

Fig. 3.26 The propagation delay can be reduced by reducing the parasitic capacitance due to the deep n^+ drain/source to p-well and the deep p^+ drain/source to n-well according to Kim *et al.*[65] The parasitic junction capacitance can be reduced by adding a graded S/D implant after the deep S/D implant. For n-channel MOS transistors, this was done by implanting phosphorus after an arsenic implant; for p-channel transistors, this was done by implanting boron at a higher energy after a boron implant at a lower energy. (Modified from Fig. 11 in H. K. Kim, S. Y. Ong, E. Quek and S. Chu, "High performance device design through parasitic junction capacitance reduction and junction leakage current suppression beyond 0.1 μm technology", *Jpn. J. Appl. Phys.*, vol. 42, no. 4B (April 2003), pp. 2144–2148.)

Shishiguchi in 2006.[66] It was further discussed by Tan *et al.*[67] in 2008. As shown in Fig. 3.27, there is a reduction in the junction capacitance due to the application of carbon co-implanted halo according to Tan *et al.*[67] In addition, as shown in Fig. 3.28 and Fig. 3.29, there is a noticeable improvement in the on current and a noticeable reduction in the propagation delay due to the application of carbon co-implanted halo according to Tan *et al.*[67]

3.5.1.2 Gate engineering and well/channel engineering

The overlap capacitance between the gate and the drain or the source can be reduced by the poly re-oxidation process. This was,

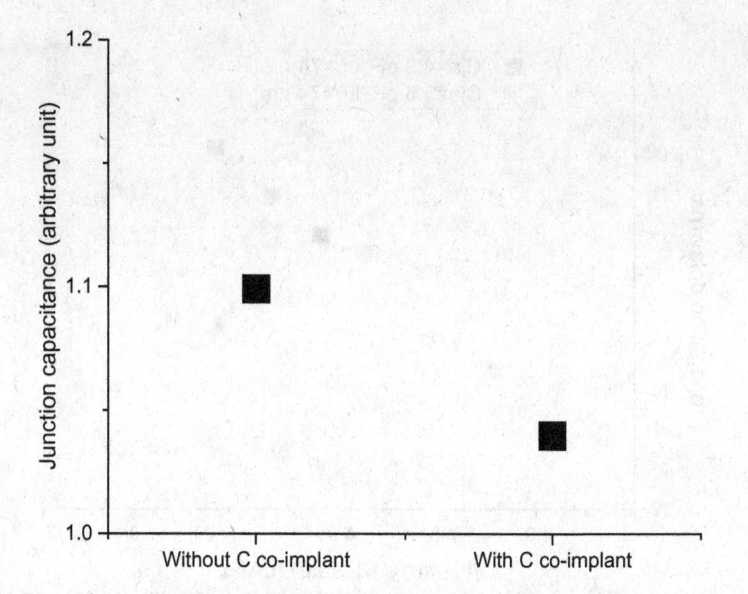

Fig. 3.27 The reduction of junction capacitance due to carbon co-implanted halo according to Tan *et al.*[67] (Modified from Fig. 15 in C. F. Tan, J. G. Lee, L. W. Teo, C. Yin, G. Lin, E. Quek and S. Chu, "A carbon co-implantation technique for formation of steep halo for nFET short channel effect improvement and performance boost", *Symp. VLSI Tech.*, pp. 32–33 (2008).)

for example, discussed by Hu *et al.*[68] As discussed above, the propagation delay can be reduced by improving the poly re-oxidation, as shown in Fig. 3.2.

Well/channel engineering is important for the reduction of parasitic capacitance. For example, Imai *et al.*[69] used a single well/local channel design for high-speed n-channel and p-channel transistors and a twin well design for low-power n-channel and p-channel transistors. As shown in the drawing on the left hand side of Fig. 3.30, the parasitic drain capacitance for high-speed n-channel and p-channel transistors can be reduced. In addition, as shown in the drawing on the right hand side of Fig. 3.30, a mask can be used such that the threshold adjust implant and the anti-punchthrough implant end up in the channel region only such that the drain and the source are not doped by these 2 implants. It can be imagined that the drain/source parasitic capacitance can be reduced by such an approach. However,

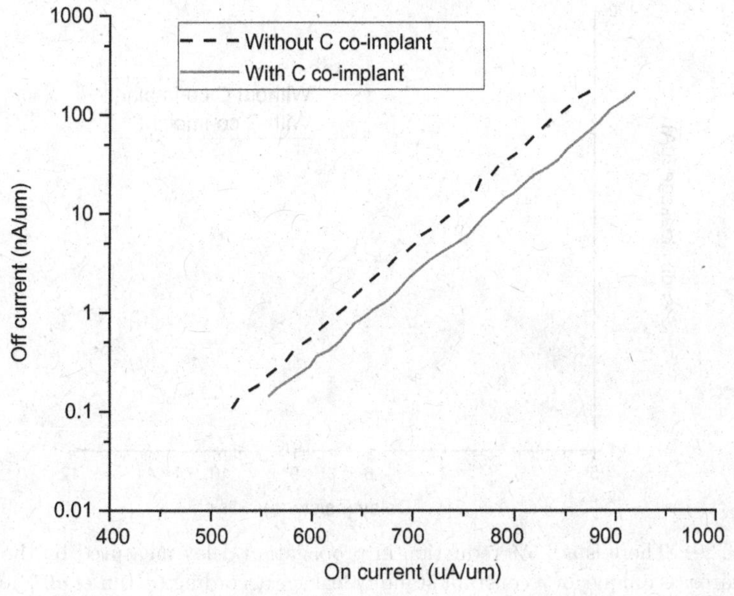

Fig. 3.28 Semi-log plot of off current versus on current for VDD = 1.0 V, showing that there is an improvement in on current for the same off current due to carbon co-implanted halo according to Tan *et al.*[67] (Modified from Fig. 6 in C. F. Tan, J. G. Lee, L. W. Teo, C. Yin, G. Lin, E. Quek and S. Chu, "A carbon co-implantation technique for formation of steep halo for nFET short channel effect improvement and performance boost", *Symp. VLSI Tech.*, pp. 32–33 (2008).)

this will increase the complexity of the CMOS process, resulting in higher cost.

3.5.1.3 SOI technology

SOI stands for silicon on insulator. Parasitic capacitance can be significantly reduced by switching from bulk CMOS technology to SOI CMOS technology. As shown in Table 3.2, there is an obvious improvement in the switching speed by using SOI CMOS technology. The data shown in Table 3.2 come from Shahidi 2002.[70]

Shahidi[70] pointed out that in general there is a 20–35% performance improvement by switching from bulk CMOS technology to SOI CMOS technology.

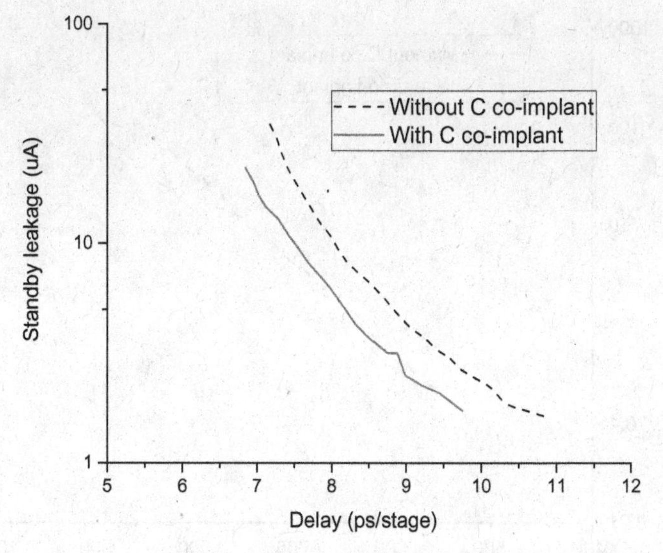

Fig. 3.29 There is a 6.5% reduction in propagation delay measured by the ring oscillator technique for a constant standby leakage according to Tan *et al.*[67] (Modified from Fig. 17 in C. F. Tan, J. G. Lee, L. W. Teo, C. Yin, G. Lin, E. Quek and S. Chu, "A carbon co-implantation technique for formation of steep halo for nFET short channel effect improvement and performance boost", *Symp. VLSI Tech.*, pp. 32–33 (2008).)

3.5.2 *Back-end techniques used to reduce parasitic capacitance*

It can be easily imagined that if the parasitic capacitance due to the parasitic capacitance between the metal lines used for interconnect is reduced, CMOS switching speed can be enhanced. The insulator used between metal lines is known as IMD (inter-metal-dielectric); it is also known as inter-layer-dielectric (ILD) or inter-level-dielectric (ILD). Thus the key point is to reduce the dielectric constant of the IMD. The dielectric constant of thermal silicon dioxide is 3.9. However, the silicon dioxide film deposited by plasma enhanced chemical vapor deposition (PECVD) for the back end can have a slightly higher dielectric constant of about 4. The dielectric constant of PECVD silicon dioxide can be reduced to below 4 by adding fluorine to 3.2–3.6.[71] Subsequently, newer IMD material has been developed with lower dielectric constant. This class of material is known as low-k dielectric. For example, SiLK™ is a low-k dielectric

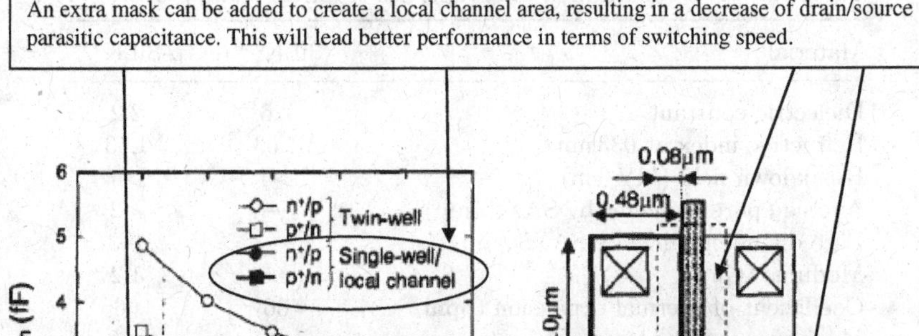

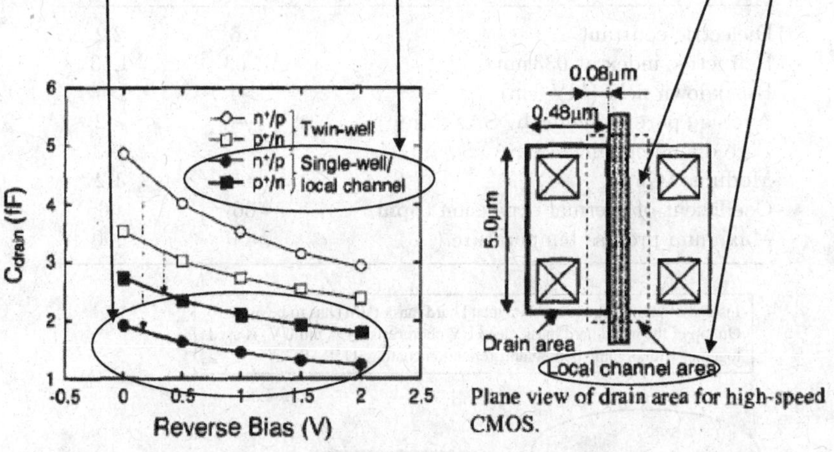

Fig. 3.30 Reduction of parasitic capacitance for high-speed n-channel and p-channel transistors by using a single well/local channel design according to Imai *et al.*[69] (Modified from Fig. 3 in K. Imai, K. Yamaguchi, N. Kimizuka, H. Onishi, T. Kudo, A. Ono, K. Noda, Y. Goto, H. Fujii, M. Ikeda, K. Kazama, S. Maruyama, T. Kuwata and T. Horiuchi, "A 0.13-μm CMOS technology integrating high-speed and low-power/high-density devices with two different well/channel structures", *IEDM Technical Digest*, (1999), pp. 667–670.)

Table 3.2. Core clock frequency of IBM 64-bit PowerPC on bulk and SOI CMOS technology (0.22 μm and 0.18 μm generations).

	Bulk 0.22 μm	SOI 0.22 μm	SOI 0.18 μm
Core clock frequency (MHz)	450	550	660

material developed by Dow Chemical; the dielectric constant can be as low as 2.6, as shown in Table 3.3. Subsequently, a newer "porous" material known as p-SiLK^{TM} developed also by Dow Chemical has even lower dielectric constant and can be considered as an ultra-low-k (ULK) material; the dielectric constant can be as low as 2.2, as shown in Table 3.3. A process for the integration of ULK dielectric material and copper is shown in Fig. 3.31, according to Richard *et al.*[72] Fig. 3.32 and Fig. 3.33 show the advantages of ULK dielectric over the more traditional low-k dielectric.

Table 3.3. A comparison of the properties of SiLK^{TM} and p-SiLK^{TM}.

Material	SiLK^{TM}	p-SiLK^{TM}
Dielectric constant	2.6	2.2
Refractive index at 633 nm	1.63	1.53
Breakdown field (MV/cm)	>4	>4
Average pore diameter by SAXS (nm)	—	<2
Pore diameter range by SAXS (nm)	—	1–3
Modulus (GPa)	2.5	1.2
Coefficient of thermal expansion (ppm/°C)	66	66
Maximum process temperature (°C)	400	400

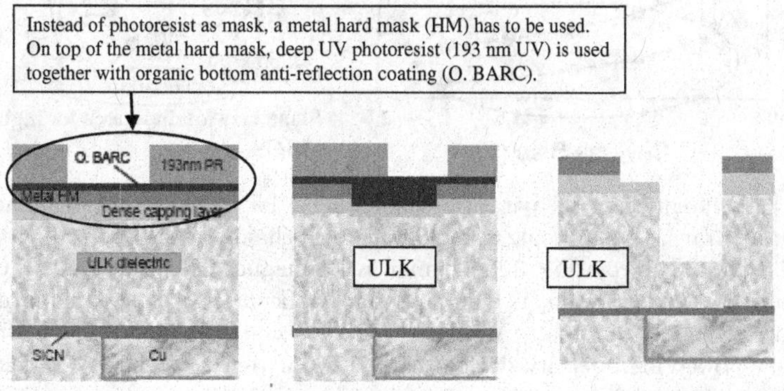

(a) Line lithography with organic BARC (b) HM etch and via lithography (c) Partial via etch and resist removal

(d) Line etch and SiCN open (e) Metal deposition and optimized CMP

Fig. 3.31 Trench First Hard Mask architecture optimized for low-k and ULK integration; via etch, resist removal, line etch, SiCN open [steps c) and d) above] are performed in a single operation according to Richard *et al.*[72] (Modified from Fig. 3 in E. Richard, R. Fox, C. Monget, M. Zaleski, P. Ferreira, A. Guvenilir, P. Brun, E. Oilier, M. Guillermet, M. Mellier, S. Petitdidier, R. Delsol, W. Besling, L. Marinier, G. Imbert, A. Lagha, L. Broussous, M. Rasco, C. Cregut, S. Downey, G. Huang, M. Haond, N. Cave and A. Perera, "Manufacturability and speed performance demonstration of porous ULK (k = 2.5) for a 45 nm CMOS platform", *Symp. VLSI Tech.*, pp. 178–179, 2007.)

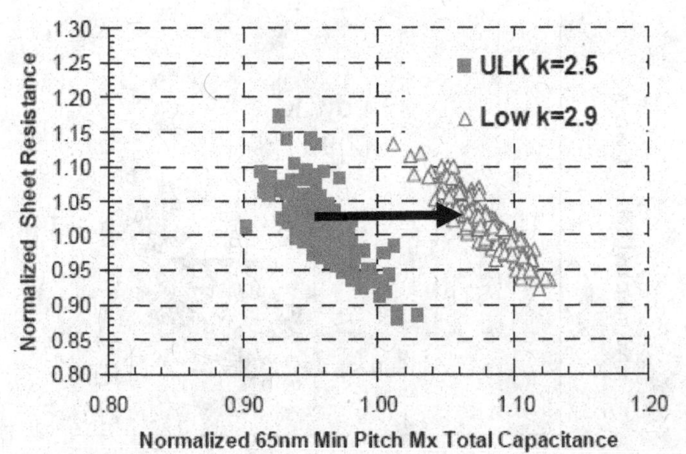

Fig. 3.32 Normalized sheet resistance vs. normalized capacitance when a low-k dielectric with a dielectric constant K = 2.9 is replaced by an ultra-low-k (ULK) material with K = 2.5, showing the significant advantage of ULK compared to low-k according to Richard *et al.*[72] (Note: The black arrow indicates that the parasitic capacitance is significantly decreased by migration to ULK technology.) (Modified from Fig. 6 in E. Richard, R. Fox, C. Monget, M. Zaleski, P. Ferreira, A. Guvenilir, P. Brun, E. Oilier, M. Guillermet, M. Mellier, S. Petitdidier, R. Delsol, W. Besling, L. Marinier, G. Imbert, A. Lagha, L. Broussous, M. Rasco, C. Cregut, S. Downey, G. Huang, M. Haond, N. Cave and A. Perera, "Manufacturability and speed performance demonstration of porous ULK (k = 2.5) for a 45 nm CMOS platform", *Symp. VLSI Tech.*, pp. 178–179, 2007.)

An overall speed performance enhancement of >10% has been confirmed by Richard *et al.* in 2007 using a microprocessor fabricated by 65 nm CMOS technology when a low-k dielectric with a dielectric constant K = 2.9 is replaced by an ultra-low-k (ULK) material with K = 2.5. However the mechanical strength of such a system has to be improved such that it can be used in practical manufacturing.

As shown in Table 3.3, the introduction of porosity can lead to significantly lower dielectric constant. The mechanical properties of ULK dielectric material are also important; for example, Wang *et al.* reported their study on the mechanical properties of ULK dielectric materials.[73] As shown in Table 3.3, Young's modulus is also significantly reduced, implying that the mechanical strength is lowered such that there may be problem for large scale production. For example, there is shear stress due to chemical mechanical polishing which can be a problem especially for ULK material.[74,75] (Note: CMP is used

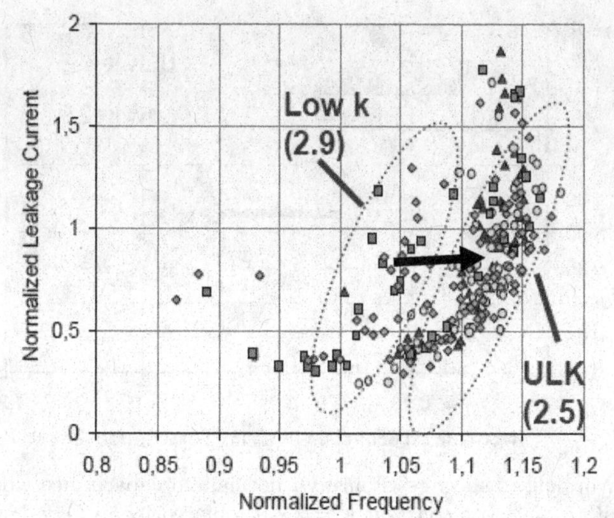

Fig. 3.33 Normalized leakage current vs. normalized frequency in a microprocessor which uses 9 metal levels, showing the significant advantage of ULK compared to low-k according to Richard *et al.*[72] (Note: The black arrow indicates that the allowed clock frequency of microprocessor is significantly increased by migration to ULK technology.) (Modified from Fig. 7 in E. Richard, R. Fox, C. Monget, M. Zaleski, P. Ferreira, A. Guvenilir, P. Brun, E. Oilier, M. Guillermet, M. Mellier, S. Petitdidier, R. Delsol, W. Besling, L. Marinier, G. Imbert, A. Lagha, L. Broussous, M. Rasco, C. Cregut, S. Downey, G. Huang, M. Haond, N. Cave and A. Perera, "Manufacturability and speed performance demonstration of porous ULK (k = 2.5) for a 45 nm CMOS platform", *Symp. VLSI Tech.*, pp. 178–179, 2007.)

during the copper/low-k back end process, as shown in Fig. 3.31.) To make practical manufacturing possible, it is important to reduce the mechanical stress during chemical mechanical polishing (CMP). In the Jan. 2005 issue of IEEE Spectrum, Brown introduced a new CMP technology invented by Applied Materials.[76] This is known as electrochemical mechanical polishing (ECMP). Subsequently, ECMP has been intensively studied and discussed in the literature.[77–80]

3.6 CMOS Technology Improvement by Using High-k Dielectric and Metal Gate

As seen in Eqs. (3.3)–(3.9), the drain current can be increased by increasing Cox. This can be achieved by using thinner gate oxide. The tradition gate dielectric material is silicon dioxide. It can be scaled down all the way to a physical thickness of about 1.6 nm. Further scaling down the thickness of silicon dioxide can lead to strong gate leakage current. Further increase of Cox can be achieved by using a high-k dielectric material as the gate dielectric.

Another issue is polysilicon depletion. There is a very thin depletion layer formed in the polysilicon at the interface between the polysilicon and the gate dielectric, resulting in a smaller Cox. This problem can be solved by using a metal gate instead of a polysilicon gate.

In the traditional process flow for a CMOS process based on the polysilicon gate and silicon dioxide gate dielectric, there is a short high temperature annealing step at about 1000°C to activate the implanted dopants. The metal gate/high-k dielectric combination may or may not be able to stand this high temperature. This leads to 2 different approaches: gate-first and gate-last. For the gate-first approach, the metal gate/high-k dielectric combination is done before this short high temperature annealing step at about 1000°C. For the gate-last approach, the metal gate/high-k dielectric combination is done before this short high temperature annealing step at about 1000°C. A comparison of these 2 approaches has been discussed by various authors, for example, by Hoffmann.[81] Recently, the gate-last approach appears to be the better approach.

An important issue is the presence of an interfacial oxide between high-k dielectric and silicon. Since the dielectric constant of silicon dioxide is only 3.9, even the presence of an ultrathin interfacial oxide layer can make the high-k dielectric much less effective. According to a study by Lau,[82] the growth of the interfacial oxide between high-k dielectric and silicon is not a simple process. In fact, it involves two contradictory effects. One effect is the oxidation of silicon, resulting in an increase in the thickness of the interfacial oxide. The other effect is the SiO volatilization, resulting in a decrease in the thickness of the interfacial oxide. Experimental evidence of the second effect is shown in Fig. 3.34. (Note: Silicon monoxide with the chemical formula of SiO can sublime at relatively low temperatures.) Similarly, the growth of the interfacial oxide between high-k dielectric and germanium is not a simple process. In fact, it involves two contradictory effects. One effect is the oxidation of germanium, resulting in an increase in the thickness of the interfacial oxide. The other effect is the SiO

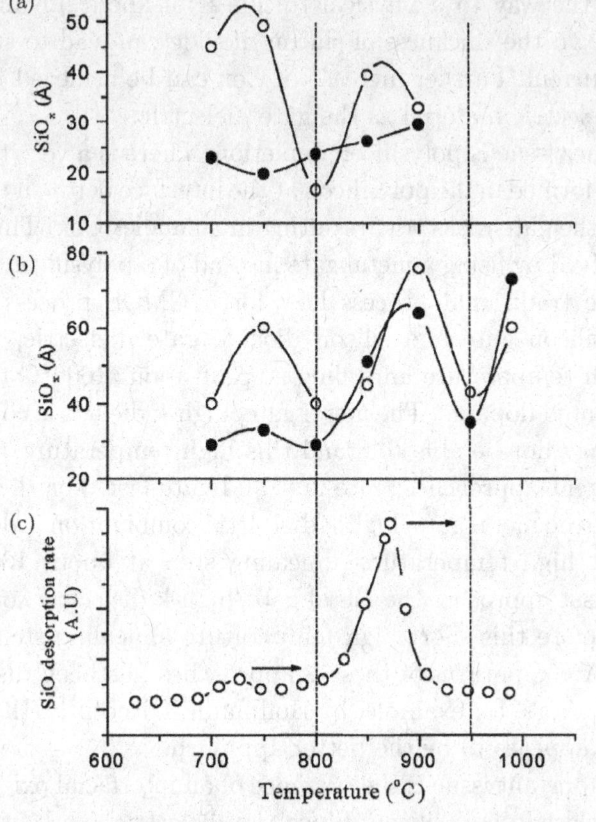

Fig. 3.34 (a) The thickness of silicon dioxide interfacial layer at the Ta_2O5/n^+-Si interface for Ta_2O_5/n^+-Si samples as a function of RTP temperature for N_2O (open circles) and O_2 (closed circles) RTP for 30 s. (b) The thickness of silicon dioxide film on Si for bare n^+-Si samples as a function of RTP temperature for N_2O (open circles) and O_2 (closed circles) RTP for 30 s. (c) The silicon monoxide SiO desorption rate as a function of temperature in UHV. (This figure comes from the work of the author.[82])

volatilization, resulting in a decrease in the thickness of the interfacial oxide. (Note: Germanium monoxide with the chemical formula of GeO can sublime at relatively low temperatures.) It appeared to the author that GeO volatilization can be more readily observed then SiO volatilization. For example, in 2008, Kita *et al.* published a good paper on their evidence of GeO volatilization from GeO_2/Ge.[83]

The most effective approach to remove this detrimental interfacial oxide appears to be using a sacrificial metal film to getter the oxygen away from the interfacial oxide according to Kim *et al.*[84] Subsequently, this approach has been confirmed by various authors. For example, Choi *et al.* tested and confirmed the effectiveness of this approach.[85,86] Using this approach, Choi *et al.* achieved an EOT as thin as 0.5 nm. However, as shown in Fig. 3.35, the penalty was a lower electron mobility. This approach was also discussed by Ando.[87] Ando pointed out that this approach can achieve "zero" interfacial oxide; however, device "reliability" may be degraded. Thus a little bit of interfacial oxide may be good for device reliability.

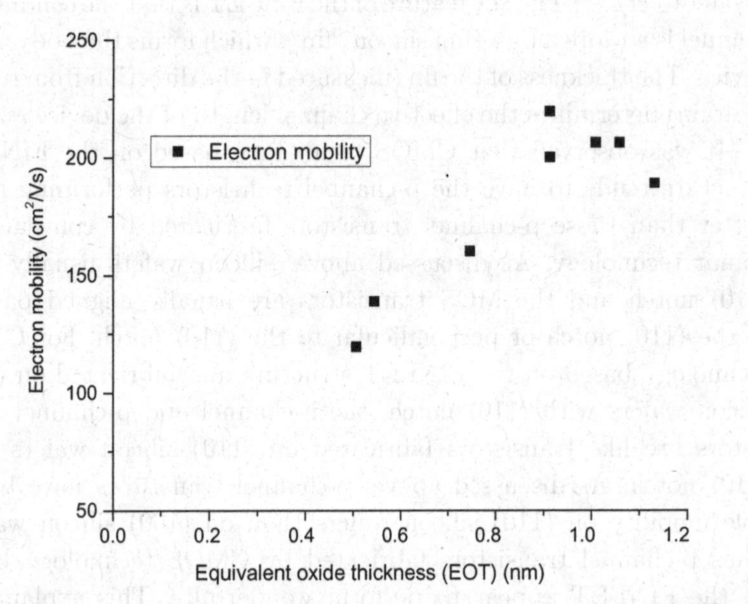

Fig. 3.35 High field electron mobility degradation as a function of equivalent oxide thickness (EOT). The mobility values were taken at Eeff = 1 MV/cm. EOT shown in this figure was actually approximately equal to EOT at inversion subtracted by about 0.4 nm. This figure originates from Choi *et al.*[85] (Modified from Fig. 11 in K. Choi, H. Jagannathan, C. Choi, L. Edge, T. Ando, M. Frank, P. Jamison, M. Wang, E. Cartier, S. Zafar, J. Bruley, A. Kerber, B. Linder, A. Callegari, Q. Yang, S. Brown, J. Stathis, J. Iacoponi, V. Paruchuri and V. Narayanan, "Extremely scaled gate-first high-k/metal gate stack with EOT of 0.55 nm using novel interfacial layer scavenging techniques for 22nm technology node and beyond", *Symp. VLSI Tech.*, pp. 138–139, 2009.)

3.7 CMOS Technology Improvement by Using 3-D Device Structure Like FINFET

For integrated circuits, “packing density” is also important. Scaling down the size of devices can increase packing density. The adoption of three-dimensional (3-D) device structures like the FINFET structure will help to increase packing density.

The term FinFET was coined by University of California, Berkeley researchers (Profs. Chenming Hu, Tsu-Jae King-Liu and Jeffrey Bokor) to describe a nonplanar, double-gate transistor built on an SOI substrate,[88] based on the earlier DELTA (single-gate) transistor design by Hisamoto *et al.*[89] and the earlier folded-channel MOSFET by Hisamoto *et al.*[90] The key feature of the FinFET is that the conducting channel is wrapped by a thin silicon “fin”, which forms the body of the device. The thickness of the fin (measured in the direction from source to drain) determines the effective channel length of the device.

It was observed that CMOS technology based on the FINFET structure tends to have the p-channel transistors performing much better than those p-channel transistors fabricated by conventional planar technology. As discussed above, silicon wafers usually have ⟨110⟩ notch and the MOS transistors are usually aligned parallel to the ⟨110⟩ notch or perpendicular to the ⟨110⟩ notch. For CMOS technology based on the FINFET structure and fabricated on (100) silicon wafers with ⟨110⟩ notch, the n-channel and p-channel transistors are like transistors fabricated on (110) silicon wafers with ⟨110⟩ notch. As discussed above, p-channel transistors have better hole mobility on (110) silicon wafers than on (100) silicon wafers. Thus p-channel transistors fabricated by CMOS technology based on the FINFET appear to perform wonderfully. This explanation has been verified as follows. In 2012, Young *et al.*[91] reported that p-channel transistors fabricated by FINFET technology on (100) silicon wafers with ⟨110⟩ notch do perform very well but p-channel transistors fabricated by FINFET technology on (100) silicon wafers with ⟨100⟩ notch do not enjoy the advantage discussed above.

As discussed above, for (100) Si wafer with ⟨110⟩ notch, the n-channel and p-channel transistors are like transistors fabricated on (110) silicon wafers with ⟨110⟩ notch. Figure 3.36 and Fig. 3.37

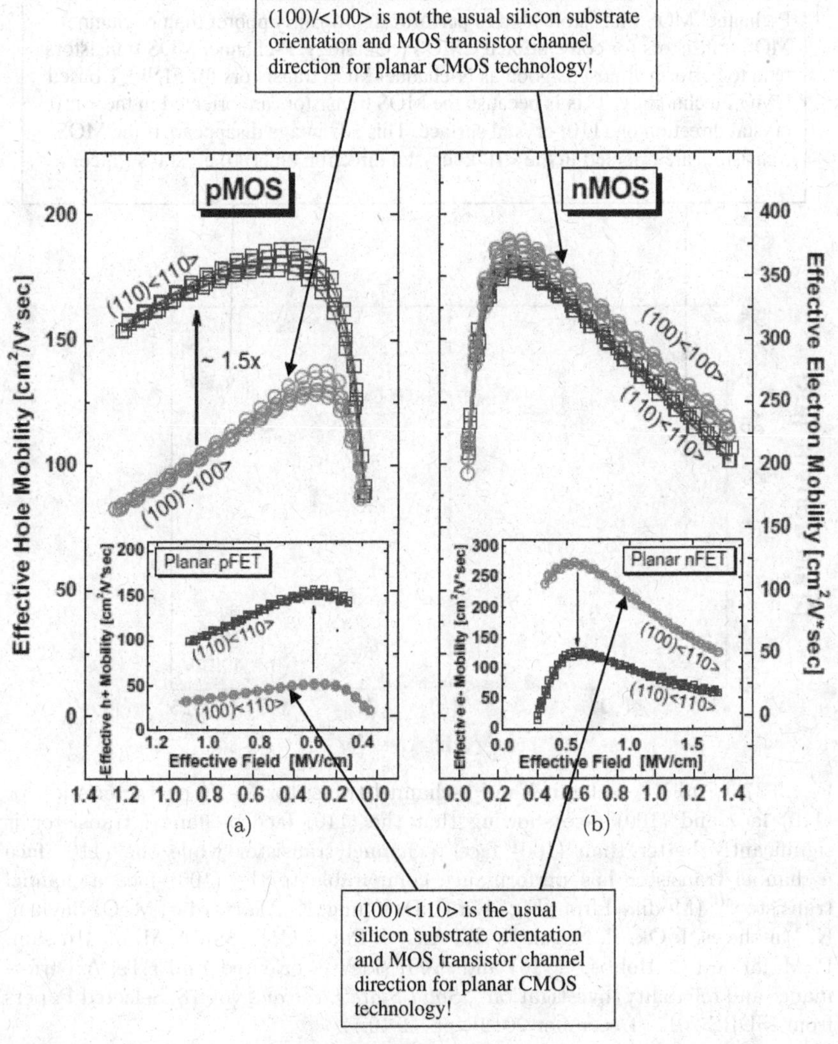

Fig. 3.36 FinFET effective electron mobility and hole mobility for (110) face and (100) face showing that the (110) face p-channel transistor is significantly better than (100) face p-channel transistor while the (110) face n-channel transistor has performance comparable to the (100) face n-channel transistor according to Young *et al.*[91] (Modified from Fig. 2 in C.D. Young, K. Akarvardar, M.O. Baykan, K. Matthews, I. Ok, T. Nagai, K.-W. Ang, J. Pater, C.E. Smith, M.M. Hussain, P. Majhi and C. Hobbs, "(110) and (100) sidewall-oriented FinFETs: A performance and reliability investigation", Solid-State Electron., vol. 78, Selected Papers from ISDRS 2011 (December 2012), pp. 2–10.)

P-channel MOS transistors tend to perform significantly poorer than N-channel MOS transistors for conventional CMOS technology. P-channel MOS transistors tend to perform almost as good as N-channel MOS transistors for FINFET based CMOS technology. This is because the MOS transistors are oriented in the <110> crystal direction on (110) crystal surface. This advantage disappears if the MOS transistors are oriented in the <100> crystal direction on (100) crystal surface.

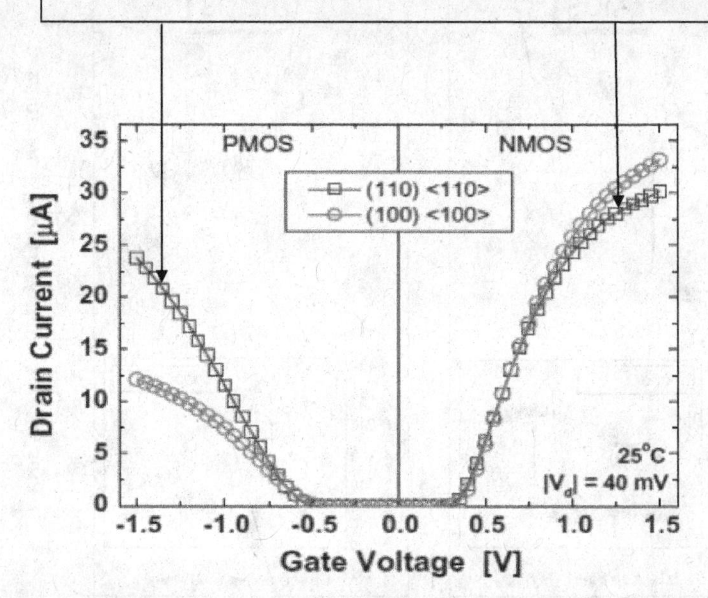

Fig. 3.37 FinFET n-channel and p-channel transistor Id-Vg characteristics for (110) face and (100) face showing that the (110) face p-channel transistor is significantly better than (100) face p-channel transistor while the (110) face n-channel transistor has performance comparable to the (100) face n-channel transistor.[91] (Modified from Fig. 4 in C. D. Young, K. Akarvardar, M. O. Baykan, K. Matthews, I. Ok, T. Nagai, K.-W. Ang, J. Pater, C. E. Smith, M. M. Hussain, P. Majhi and C. Hobbs, "(110) and (100) sidewall-oriented FinFETs: A performance and reliability investigation", *Solid-State Electron.*, vol. 78, Selected Papers from ISDRS 2011 (December 2012), pp. 2–10.)

show that the p-channel FINFET performs wonderfully; however, this advantage will disappear for (100) Si wafer with ⟨100⟩ notch.

3.8 CMOS Technology Improvement by Using Materials with Higher Mobility than Silicon

As discussed above, it is expected that CMOS technology based on semiconductor materials with higher mobility than silicon will have the advantage of higher drive current and thus higher switching speed.

Table 3.4. Electron and hole mobilities in Si, Ge, GaAs, InGaAs and InAs according to Lubow *et al.*[92]

Material	Si	Ge	GaAs	InGaAs	InAs
Electron mobility ($cm^2V^{-1}s^{-1}$)	1350	3600	8000	11200	30000
Hole mobility ($cm^2V^{-1}s^{-1}$)	480	1800	300	300	450

Table 3.4 is a table of electron and hole mobilities of various semiconductor materials according to Lubow *et al.*[92] As shown in Table 3.4, it is well known that germanium (Ge) has both higher electron mobility and also higher hole mobility compared to silicon. However, the advantage of Ge over Si has been demonstrated for p-channel transistors. For n-channel transistors, the poor Ohmic contact on n^+-Ge becomes the limiting factor. Subsequently, a lot of work has been done to improve the Ohmic contact on n^+-Ge. A review on Ge-based nano devices has been given by Pillarisetty.[93] Besides Ge, it is known that some III-V semiconductors have higher electron mobility than silicon. For example, as shown in Table 3.4, GaAs has larger electron mobility compared to both Si and Ge; however, GaAs has smaller hole mobility compared to both Si and Ge. Recently, some III-V semiconductors have been demonstrated to have larger hole mobility compared to both Si and Ge.

As discussed above, it is easier to fabricate good p-channel MOS transistors on Ge than n-channel MOS transistors on Ge. As shown in Fig. 3.38, Fig. 3.39 and Fig. 3.40, Gupta *et al.*[94] pointed out that switching the channel orientation from $\langle 110 \rangle$ to $\langle 100 \rangle$ can improve p-channel MOS transistors on Ge just like p-channel transistors on Si.

Fang *et al.* fabricated InAsSb-on-insulator MOS transistors and reported their results in 2012.[95] Figure 3.41 shows that InAsSb has high electron mobility. Addition of Sb to InAs increases the electron mobility to about 4000 $cm^2V^{-1}s^{-1}$. It is known that Ge has an electron mobility with the value of 3900 $cm^2V^{-1}s^{-1}$. (Note: This value comes from Appendix G in the book *Physics of Semiconductor Devices* by S. M. Sze, second edition, 1981. Table 3.4 shows a value of 3600 $cm^2V^{-1}s^{-1}$.) The addition of Sb to InAs increases the electron mobility to about 4000 $cm^2V^{-1}s^{-1}$, which apparently seems to be about the same as the value of the electron mobility of

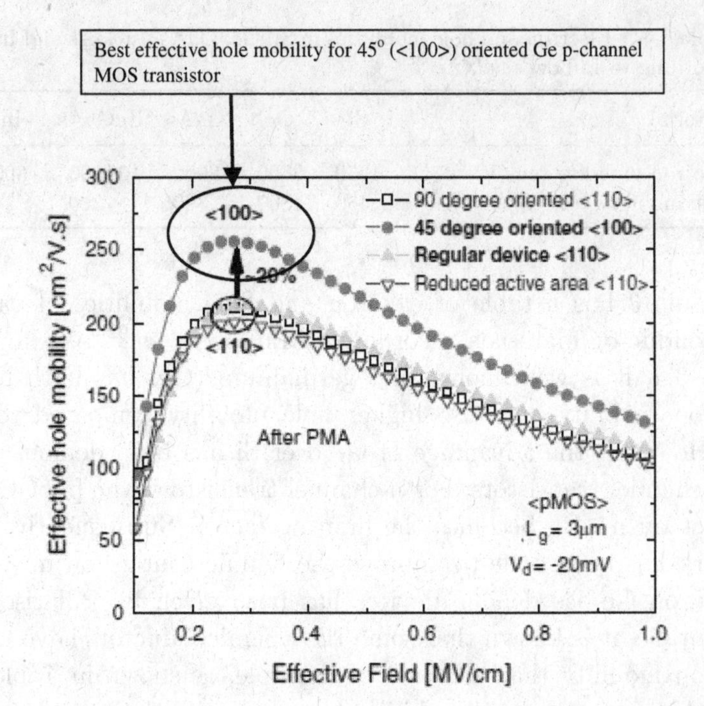

Fig. 3.38 Effective hole mobility measured as a function of the effective field for all the available devices showing optimum performance for 45° (⟨100⟩) oriented device according to Gupta *et al.* 2010.[94] (Modified from Fig. 8 in S.D. Gupta, J. Mitard, G. Eneman, B. De Jaeger, M. Meuris and M. M. Heyns, "Performance enhancement in Ge pMOSFETs with ⟨100⟩ orientation fabricated with a Si-compatible process flow", *Microelectronic Engineering*, vol. 87, no. 11 (Nov. 2010), pp. 2115–2118.)

germanium. The readers should note that the electron mobility value of 3900 $cm^2V^{-1}s^{-1}$ from the book by S. M. Sze is for bulk germanium with a low doping concentration. It can be imagined that the electron mobility should be significantly smaller at the interface of the gate dielectric and germanium. In addition, in a practical germanium MOS transistor, germanium has to be doped for threshold voltage control, suppression of punchthrough, etc. Figure 3.42 shows the effective electron mobility in practical Ge MOS transistor according to Toriumi *et al.*[96] Figure 3.42 shows that the peak electron mobility is about 800 $cm^2V^{-1}s^{-1}$ for practical Ge MOS transistors. Thus the increase of the electron mobility to about 4000 $cm^2V^{-1}s^{-1}$ by

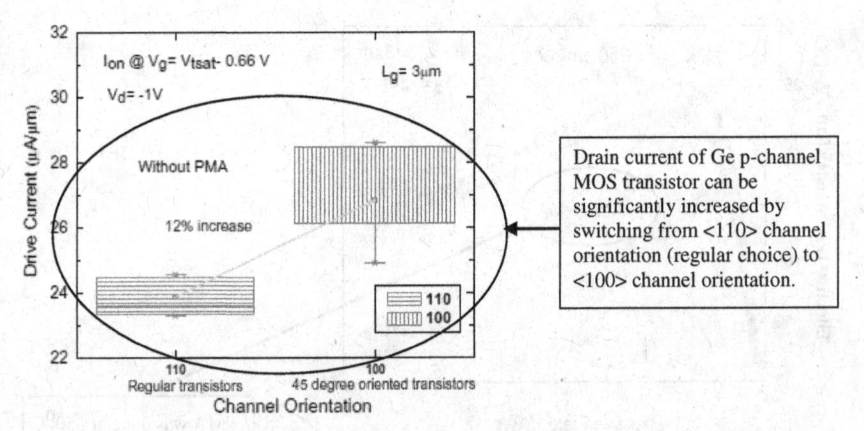

Fig. 3.39 Drive current enhancement with change in orientation of the Ge p-channel MOS transistor from ⟨110⟩ channel orientation to ⟨100⟩ channel orientation according to Gupta *et al.* 2010.[94] (Modified from Fig. 7 in S.D. Gupta, J. Mitard, G. Eneman, B. De Jaeger, M. Meuris and M. M. Heyns, "Performance enhancement in Ge pMOSFETs with ⟨100⟩ orientation fabricated with a Si-compatible process flow", *Microelectronic Engineering*, vol. 87, no. 11 (Nov. 2010), pp. 2115–2118.)

adding Sb to InAs at the interface of the gate dielectric and InAsSb is a significant technological improvement.

Nainani *et al.* showed that InGaSb has larger hole mobility compared to Ge.[97] Figure 3.43 shows that the peak hole mobility is about 900 $cm^2V^{-1}s^{-1}$ for InGaSb with 1.7% biaxial compression. The value of about 900 $cm^2V^{-1}s^{-1}$ may not look impressive because it is known that Ge has a hole mobility with the value of 1900 $cm^2V^{-1}s^{-1}$ according to the book by S.M. Sze or the value of 1800 cm $^2V^{-1}s^{-1}$ according to Table 3.4. The above 2 numbers are for bulk germanium with a low doping concentration. The numbers should be different for the case of an MOS transistor involving germanium. It can be imagined that the hole mobility should be smaller at the interface of the gate dielectric and germanium. In addition, in a practical germanium MISFET, germanium has to be doped for threshold voltage control, suppression of punchthrough, etc. Figure 3.38 and Fig. 3.40 show more reasonable values for the hole mobility in practical Ge MOS transistors according to Gupta *et al.* Comparing with the values of hole mobility from Gupta *et al.*, then the peak hole mobility with

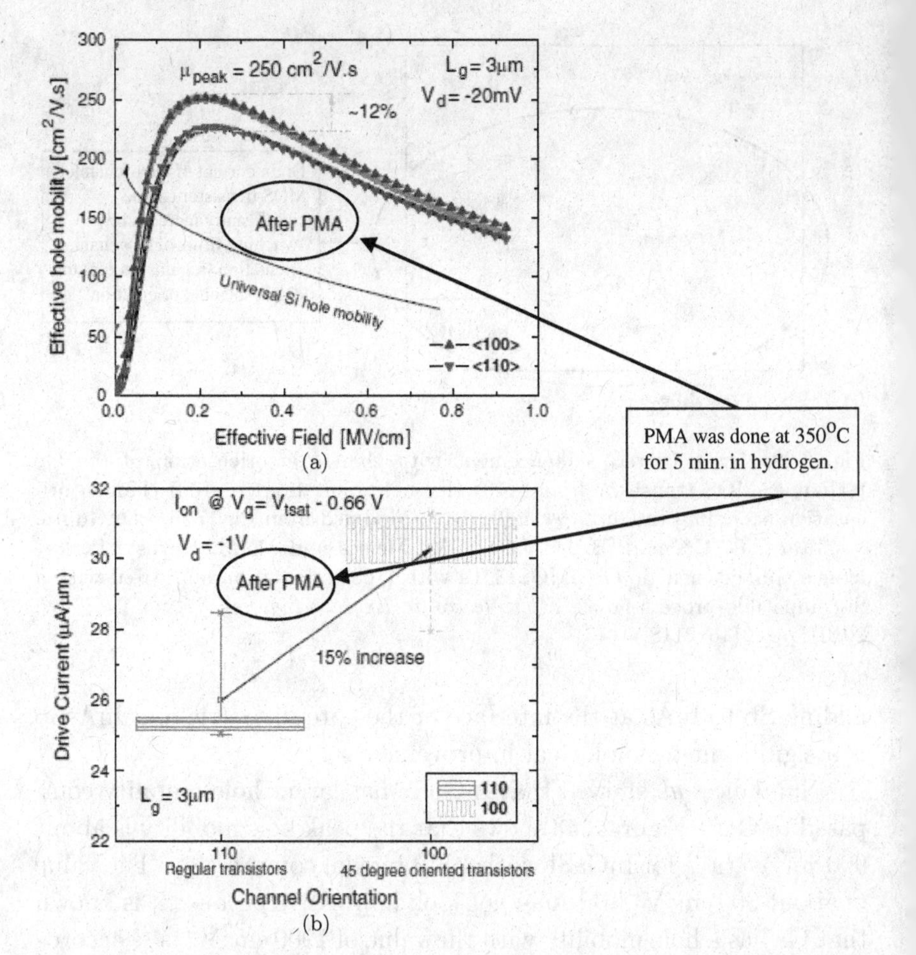

Fig. 3.40 (a) Effective hole mobility (at low $V_d = -20\,\text{mV}$) measured as a function of effective field for ⟨110⟩ and ⟨100⟩ oriented devices shows 12% enhancement after PMA. (b) The magnitude of the drive current (at high $V_d = -1\,\text{V}$) enhancement with change of channel orientation after PMA according to Gupta *et al.* 2010.[94] (Note: PMA was done at 350°C for 5 min. in hydrogen.) (Modified from Fig. 10 in S.D. Gupta, J. Mitard, G. Eneman, B. De Jaeger, M. Meuris and M.M. Heyns, "Performance enhancement in Ge pMOSFETs with ⟨100⟩ orientation fabricated with a Si-compatible process flow", *Microelectronic Engineering*, vol. 87, no. 11 (Nov. 2010), pp. 2115–2118.)

a value of about $900\,\text{cm}^2\text{V}^{-1}\text{s}^{-1}$ for InGaSb with 1.7% biaxial compression will look impressive. More discussion on better hole mobility can be found in a paper by Nainani *et al.*[98]

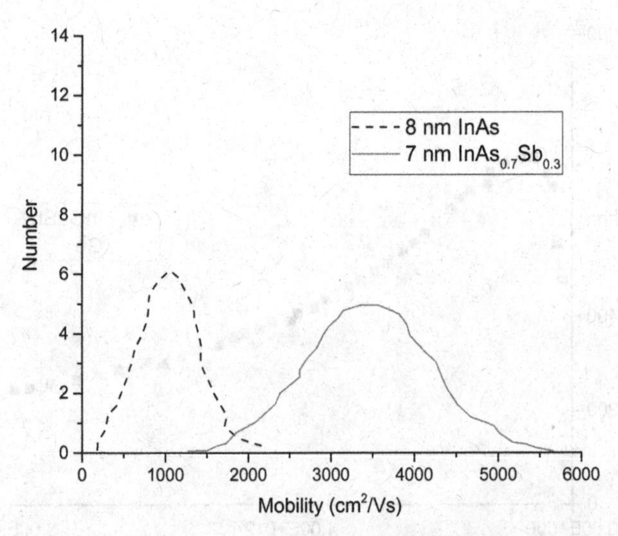

Fig. 3.41 Addition of Sb to InAs increases the electron mobility to about $4000\,cm^2V^{-1}s^{-1}$ according to Fang *et al.*[95] (Modified from Fig. 1 in H. Fang, S. Chuang, K. Takei, H. S. Kim, E. Plis, C.-H. Liu, S. Krishna, Y.-L. Chueh and A. Javey, "Ultrathin-body high-mobility InAsSb-on-insulator field-effect transistors", *IEEE Electron Dev. Lett.*, vol. 33, no. 4 (April 2012), pp. 504–506.)

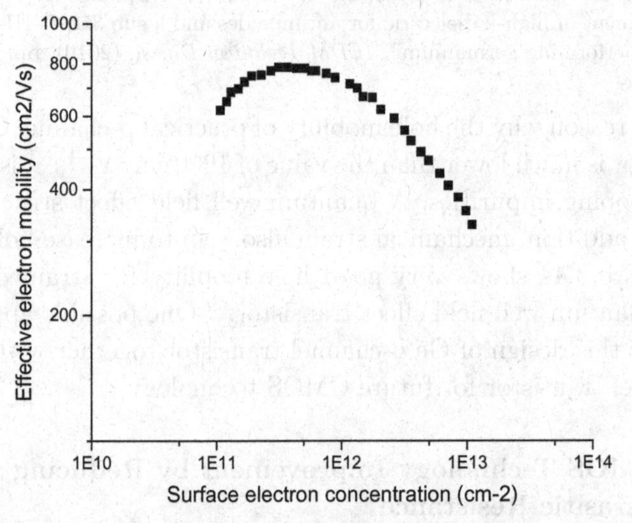

Fig. 3.42 The electron mobility in a practical MOS transistor based on germanium according to Toriumi *et al.*[96] (Modified from Fig. 10 in A. Toriumi, C. H. Lee, T. Nishimura, S. K. Wang, K. Kita and K. Nagashio, "Recent progress of Ge technology for a post-Si CMOS", *ECS Trans.*, vol. 35, no. 3 (2011), pp. 443–456.)

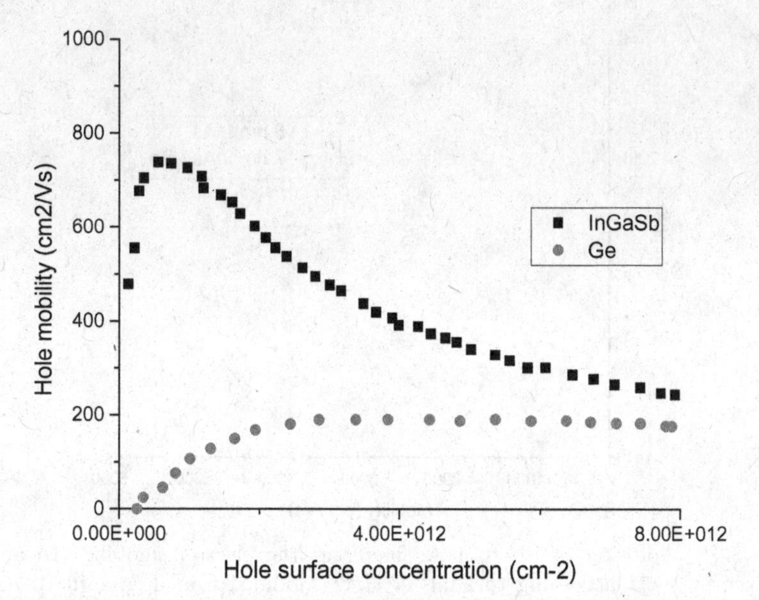

Fig. 3.43 According to Nainani *et al.* 2010,[97] $In_xGa_{1-x}Sb$ has higher hole mobility than germanium. (Note: There was 1.7% biaxial stress.) (Modified from Fig. 18(b) in A. Nainani, T. Irisawa, Z. Yuan, Y. Sun, T. Krishnamohan, M. Reason, B. R. Bennett, J. B. Boos, M. G. Ancona, Y. Nishi and K. C. Saraswat, "Development of high-k dielectric for antimonides and a sub 350°C III-V pMOSFET outperforming germanium", *IEDM Technical Digest*, (2010), pp. 138–141.)

One reason why the hole mobility of practical p-channel Ge-based transistor is much lower than the value of 1900 cm $^2V^{-1}s^{-1}$ is because of the doping impurities. A quantum well field effect structure can help. In addition, mechanical strain also help to increase hole mobility. Figure 3.44 shows very good hole mobility for strained germanium quantum well field effect transistor.[99] One possible application is to use this design of Ge p-channel transistor together with a III-V n-channel transistor for future CMOS technology.

3.9 CMOS Technology Improvement by Reducing Parasitic Resistance

Besides reducing parasitic capacitance, it is expected that higher switching speed can be achieved by the reduction of parasitic resistance. Parasitic resistance can be reduced by silicide technology. For

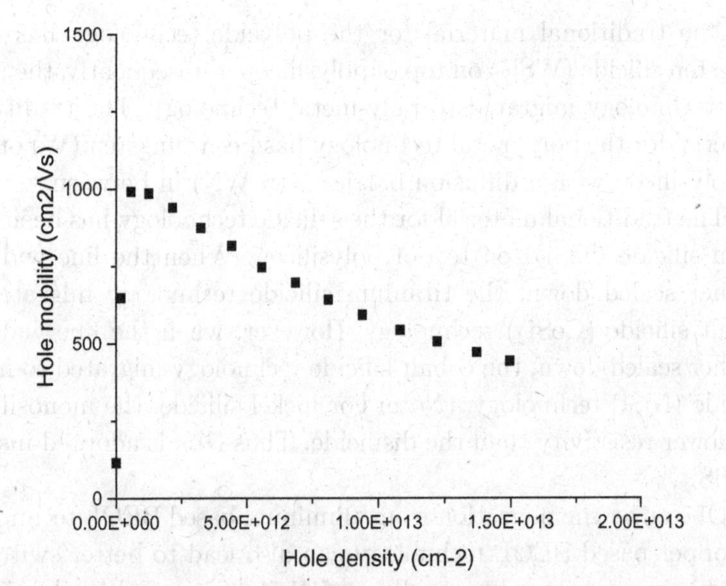

Fig. 3.44 The hole mobility in a practical MOS transistor based on germanium compared to that in a practical MOS transistor based on silicon according to Pillarisetty *et al.*[99] Please note that the p-channel transistor based on germanium uses a quantum well design. The hole mobility for a practical p-channel MOS transistor enhanced by mechanical strain is about 200 cm^2/Vs. Thus there is a 4× improvement in hole mobility. (Modified from Fig. 14 in R. Pillarisetty, B. Chu-Kung, S. Corcoran, G. Dewey, J. Kavalieros, H. Kennel, R. Kotlyar, V. Le, D. Lionberger, N. Metz, N. Mukherjee, J. Nah, W. Rachmady, M. Radosavljevic, U. Shah, S. Taft, H. Then, N. Zelick and R. Chau, "High mobility strained germanium quantum well field effect transistor as the p-channel device option for low power (Vcc = 0.5 V) III-V CMOS architecture", *IEDM Technical Digest*, (2010), pp. 150–153.)

polycide technology, only the gate parasitic resistance is reduced. For salicide (self-aligned silicide) technology, the drain and source parasitic resistances are reduced; of course, the gate parasitic resistance is also reduced. Thus, salicide technology looks superior to polycide technology. However, for example, the packing density can be a more important concern than speed; self-aligned contact (SAC) is important for achieving high packing density and the polycide technology is more compatible with SAC technology. Thus the polycide technology may be the preferred technology for DRAM instead of the silicide technology.

The traditional material for the polycide technology has been tungsten silicide (WSi_2) on top of polysilicon. Subsequently, the polycide technology migrated to poly-metal technology. The traditional material for the poly-metal technology has been tungsten (W) on top of polysilicon with a diffusion barrier (e.g. WN) in between.

The traditional material for the salicide technology has been titanium silicide ($TiSi_2$) on top of polysilicon. When the line width is further scaled down, the titanium silicide technology migrated to cobalt silicide ($CoSi_2$) technology. However, when the line width is further scaled down, the cobalt silicide technology migrated to nickel silicide (NiSi) technology. (Note: For nickel silicide, the monosilicide has lower resistivity than the disilicide. Thus NiSi is adopted instead of $NiSi_2$.)

Of course, the migration from aluminum based BEOL technology to copper based BEOL technology can also lead to better switching speed because copper has smaller resistivity compared to aluminum.

3.10 BiCMOS Technology

The switching speed is decided by the speed of charging up parasitic capacitance. Thus the increase of drive current can lead to faster switching speed. When it is necessary to drive a "long" line, a bipolar transistor is superior to an MOS transistor because of a higher drive current available from a bipolar transistor compared to an MOS transistor. Such a situation may happen at the output of digital integrated circuits. BiCMOS technology is basically using CMOS technology for the main logic portion with some bipolar output stages. Thus BiCMOS technology can sometimes lead to significant switching speed improvement.

References

[1] L. L. Vadasz, A. S. Grove, T. A. Rowe and G. E. Moore, "Silicon-gate technology", *IEEE Spectrum*, vol. 6, no. 10 (Oct. 1969), pp. 28–35.

[2] K. Shohno, K. J. Callalan and M. Hirayama, "47-stage P-MOS ring oscillator circuit using the two-photomask fabrication process", *Jpn. J. Appl. Phys.*, vol. 17, no. 6 (June 1978), pp. 1133–1134.

[3] M. Yoshimi, K. Tsuchiya, M. Iwase, M. Takahashi, E. Nishimura, T. Suzuki and Y. Kato, "Study of the operation speed of half-micron design rule CMOS ring oscillators", *Electron. Lett.*, vol. 17, no. 6 (June 1978), pp. 1133–1134.

[4] P. G. Y. Tsui, H.-H. Tseng, M. Orlowski, S.-W. Sun, P. J. Tobin, K. Reed and W. J. Taylor, "Suppression of MOSFET reverse short channel effect by N_2O gate poly reoxidation process", *IEDM Technical Digest*, (1994), pp. 501–504.

[5] M. Bhusan, A. Gattiker, M. B. Ketchen and K. K. Das, "Ring oscillators for CMOS process tuning and variability control", *IEEE Transactions on Semiconductor Manufacturing*, vol. 19, no. 1 (Feb. 2006), pp. 10–18.

[6] F. Boeuf, M. Sellier, A. Farcy and T. Skotnicki, "An evaluation of the CMOS technology roadmap from the point of view of variability, interconnects, and power dissipation", *IEEE Trans. Electron Dev.*, vol. 55, no. 6 (June 2008), pp. 1433–1440.

[7] K. K. Thornber, "Relation of drift velocity to low-field mobility and high-field saturation velocity", *J. Appl. Phys.*, vol. 51, no. 4 (April 1980), pp. 2127–2136.

[8] E. J. Ryder and W. Shockley, "Mobilities of electrons in high electric fields", *Phys. Rev.*, vol. 81, no. 1 (1 January 1951), pp. 139–140.

[9] W. Shockley, "Hot electrons in germanium and Ohm's law", *Bell System Technical Journal*, vol. 30, no. 4 (Oct. 1951), pp. 990–1034.

[10] E. J. Ryder, "Mobility of holes and electrons in high electric fields", *Phys. Rev.*, vol. 90, no. 5 (1 June 1953), pp. 766–769.

[11] J. G. Ruch, "Electron dynamics in short channel field-effect transistors", *IEEE Trans. Electron Dev.*, vol. 19, no. 5 (March 1972), pp. 652–654.

[12] S. Y. Chou, D. A. Antoniadis and H. I. Smith, "Observation of electron velocity overshoot in sub-100-nm-channel MOSFET's in silicon", *IEEE Electron Dev. Lett.*, vol. 6, no. 12 (Dec. 1985), pp. 665–667.

[13] W. S. Lau, P. Yang, J. Z. Chian, V. Ho, C. H. Loh, S. Y. Siah and L. Chan, "Drain current saturation at high drain voltage due to pinch off instead of velocity saturation in sub-100 nm metal-oxide-semiconductor transistors", *Microelectronics Reliability*, vol. 49, no. 1 (January 2009), pp. 1–7.

[14] K. Natori, "Ballistic metal-oxide-semiconductor field effect transistor", *J. Appl. Phys.*, vol. 76, no. 8 (15 October 1994), pp. 4879–4890.

[15] P. Yang, W. S. Lau, S. W. Lai, V. L. Lo, S. Y. Siah and L. Chan, "Selection of gate length and gate bias to make nanoscale metal-oxide-semiconductor transistors less sensitive to both statistical gate length variation and temperature variation", *Solid-State Electron.*, vol. 54, no. 11 (Nov. 2010), pp. 1304–1311.

[16] M. Lundstrom, "Elementary scattering theory of the Si MOSFET", *IEEE Electron Dev. Lett.*, vol. 18, no. 7 (July 1997), pp. 361–363.

[17] W. S. Lau, P. Yang, V. Ho, L. F. Toh, Y. Liu, S. Y. Siah and L. Chan, "An explanation of the dependence of the effective saturation velocity on gate voltage in sub-0.1 μm metal-oxide-semiconductor transistors by

quasi-ballistic transport theory", *Microelectronics Reliability*, vol. 48, no. 10 (October 2008), pp. 1641–1648.
[18] A. Rahman, J. Guo, S. Datta and M. S. Lundstrom, "Theory of ballistic nanotransistors", *IEEE Trans. Electron Dev.*, vol. 50, no. 9 (September 2003), pp. 1853–1864.
[19] K. Natori, "Ballistic/quasi-ballistic transport in nanoscale transistor", *Applied Surface Science*, vol. 254 (2008), pp. 6194–6198.
[20] K. Natori, T. Shimizu and T. Ikenobe, "Multi-subband effects on performance limit of nanoscale MOSFETs", *Jpn. J. Appl. Phys.*, vol. 42, Part1, no. 4B (April 2003), pp. 2063–2066.
[21] P. Yang, W. S. Lau, V. Ho, C. H. Lo, S. Y. Siah and L. Chan, "A comparison between the quasi-ballistic transport model and the conventional velocity saturation model for sub-0.1-μm MOS transistors", *Proc. IEEE EDSSC 2007*, pp. 99–102.
[22] R. Ohba and T. Mizuno, "Nonstationary electron/hole transport in sub-0.1 μm MOS devices: correlation with mobility and low-power CMOS applications", *IEEE Trans. Electron Dev.*, vol. 48, no. 2 (Feb. 2001), pp. 338–343.
[23] K. Tatsumura, M. Goto, S. Kawanaka and A. Kinoshita, "Correlation between low-field mobility and high-field carrier velocity in quasi-ballistic-transport MISFETs scaled down to Lg = 30 nm", *IEDM Technical Digest*, (2009), pp. 465–468.
[24] J. R. Hauser, "A new and improved physics-based model for MOS transistors", *IEEE Trans. Electron Dev.*, vol. 52, no. 12 (December 2005), pp. 2640–2647.
[25] C. Jeong, D. A. Antoniadis and M. S. Lundstrom, "On backscattering and mobility in nanoscale silicon MOSFETs", *IEEE Trans. Electron Dev.*, vol. 56, no. 11 (November 2009), pp. 2762–2769.
[26] S. Takagi, A. Toriumi, M. Iwase and H. Tango, "On the universality of inversion layer mobility in Si MOSFET's: Part I — Effects of substrate impurity concentration", *IEEE Trans. Electron Dev.*, vol. 41, no. 12 (Dec. 1994), pp. 2357–2362.
[27] S. Takagi, A. Toriumi, M. Iwase and H. Tango, "On the universality of inversion layer mobility in Si MOSFET's: Part II — Effects of surface orientation", *IEEE Trans. Electron Dev.*, vol. 41, no. 12 (Dec. 1994), pp. 2363–2368.
[28] C. W. Eng, W. S. Lau, D. Vigar, S. S. Tan and L. Chan, "Effective channel length measurement of metal-oxide-semiconductor transistors with pocket implant using the subthreshold current-voltage characteristics based on remote Coulomb scattering," *Appl. Phys. Lett.*, vol. 87, no. 15 (10 October 2005), pp. 153510-1–153510-3.
[29] T. Yamanaka, S. J. Fang, H.-C. Lin, J. P. Snyder and C. R. Helms, "Correlation between inversion layer mobility and surface roughness measured by AFM", *IEEE Electron Dev. Lett.*, vol. 17, no. 4 (April 1996), pp. 178–180.
[30] M. Kondo and H. Tanimoto, "An accurate Coulomb mobility model for MOS inversion layers and its application to NO-oxynitride devices", *IEEE Trans. Electron Dev.*, vol. 48, no. 2 (Feb. 2001), pp. 265–270.

[31] G. Mazzoni, A. L. Lacaita, L. M. Perron and A. Pirovano, "On surface roughness-limited mobility in highly doped n-MOSFET's", *IEEE Trans. Electron Dev.*, vol. 46, no. 7 (July 1999), pp. 1423–1428.

[32] B. Laikhtman and P. M. Solomon, "Remote phonon scattering in field-effect transistors with a high-k insulating layer", *J. Appl. Phys.*, vol. 103, no. 1 (1 January 2008), article number 014501.

[33] M.-J. Chen, S.-C. Chang, S.-J. Kuang, C.-C. Lee, W.-H. Lee, K.-H. Cheng and Y.-H. Zhan, "Temperature-dependent remote-Coulomb-limited electron mobility in n+-polysilicon ultrathin gate oxide nMOSFETs", *IEEE Trans. Electron Dev.*, vol. 58, no. 4 (Apr. 2011), pp. 1038–1044.

[34] N. Yoshikawa, T. Tomida, M. Tokuda, Q. Liu, X. Meng, S. R. Whiteley and T. Van Duzer, "Characterization of 4 K CMOS devices and circuits for hybrid Josephson-CMOS systems", *IEEE Trans. Appl. Superconductivity*, vol. 15, no. 2 (June 2005), pp. 267–271.

[35] S.-H. Hong, G.-B. Choi, R.-H. Baek, H.-S. Kang, S.-W. Jung and Y.-H. Jeong "Low-temperature performance of nanoscale MOSFET for deep-space RF applications", *IEEE Electron Dev. Lett.*, vol. 29, no. 7 (July 2008), pp. 775–777.

[36] S. Iijima, M. Yudasaka and F. Nihey, "Carbon nanotube technology", *NEC Technical Journal*, vol. 2, no. 1 (March 2007), pp. 52–56.

[37] R. S. Ruoff and D. C. Lorents, "Mechanical and thermal properties of carbon nanotubes", *Carbon*, vol. 33, no. 7 (1995), pp. 925–930.

[38] (a) J. Hone, M. Whitley, C. Piskoti and A. Zettl, "Thermal conductivity of single-walled carbon nanotubes", *Phys. Rev. B*, vol. 59, no. 4 (15 January 1999), pp. R2514–R2516. (b) J. Hone, M. Whitley and A. Zettl, "Thermal conductivity of single-walled carbon nanotubes", *Synthetic Metals*, vol. 103, no. 1–3 (June 1999), pp. 2498–2499.

[39] S. Berber, Y.-K. Kwon and D. Tomanek, "Unusually high thermal conductivity of carbon nanotubes", *Phys. Rev. Lett.*, vol. 84, no. 20 (15 May 2000), pp. 4613–4616.

[40] M. J. Biercuk, M. C. Liaguno, M. Radosavijevic, J. K. Hyun, A. T. Johnson and J. E. Fischer, "Carbon nanotube composites for thermal management", *Appl. Phys. Lett.*, vol. 80, no. 15 (15 April 2002), pp. 2767–2769.

[41] J. J. Romero, "Carbon nanotubes take the heat off chips", *IEEE Spectrum*, vol. 44, no. 12 (Dec. 2007), pp. 12–14.

[42] W. Lin, R. Zhang, K.-S. Moon and C. P. Wong, "Synthesis of high-quality vertically aligned carbon nanotubes on bulk copper substrate for thermal management", *IEEE Transactions on Advanced Packaging*, vol. 33, no. 2 (May 2010), pp. 370–376.

[43] M. Lundstrom, *Fundamentals of Carrier Transport, Second Edition*, Cambridge University Press, 2000, pp. 166–167.

[44] L. Chang, M. Ieong and M. Yang, "CMOS circuit performance enhancement by surface orientation optimization", *IEEE Trans. Electron Dev.*, vol. 51, no. 10 (Oct. 2004), pp. 1621–1627.

[45] M. Yang, V. W. C. Chan, K. K. Chan, L. Shi, D. M. Fried, J. H. Stathis, A. I. Chou, E. Gusev, J. A. Ott, L. E. Burns, M. V. Fischetti and

M. Ieong, "Hybrid orientation technology (HOT): opportunities and challenges", *IEEE Trans. Electron Dev.*, vol. 53, no. 5 (May 2006), pp. 965–978.

[46] W. S. Lau, P. Yang, T. P. Chen, S. Y. Siah and L. Chan, "Physics of electron mobility independent of channel orientation in n-channel transistors based on (100) silicon wafers and its experimental verification", *Appl. Phys. Lett.*, vol. 97, no. 13 (27 September 2010), pp. 133508-1–133508-3.

[47] J. Singh, *Physics of Semiconductors and Their Heterostructures*, McGraw-Hill, New York, 1993, pp. 163–164.

[48] H. Sayama, Y. Nishida, H. Oda, T. Oishi, S. Shimizu, T. Kunikiyo, K. Sonoda, Y. Inoue and M. Inuishi, "Effect of ⟨100⟩ channel direction for high performance SCE immune pMOSFET with less than 0.15 μm gate length", *IEDM Technical Digest*, (1999), pp. 657–660.

[49] T. Skotnicki, C. Fenouillet-Beranger, C. Gallon, F. Boeuf, S. Monfray, A. Pouydebasque, M. Szczap. A. Farcy, F. Arnaud, S. Clere, M. Sellier, A. Cathignol, J.-P. Schoellkopf, E. Perea, R. Ferrant and H. Mingam, "Innovative materials, devices and CMOS technologies for low-power mobile multimedia", *IEEE Trans. Electron Dev.*, vol. 55, no. 1 (Jan. 2008), pp. 96–130.

[50] W. S. Lau, P. Yang, V. Ho, B. K. Lim, S. Y. Siah and L. Chan, "Reduced boron lateral ion channeling in very short p-channel transistors by switching from ⟨110⟩ to ⟨100⟩ channel orientation", *Appl. Phys. Lett.*, vol. 93, no. 23 (8 December 2008), pp. 233505-1–233505-3.

[51] P. Yang, W. S. Lau, S. W. Lai, V. L. Lo, S. Y. Siah and L. Chan, "Effects of switching from ⟨110⟩ to ⟨100⟩ channel orientation and tensile stress on n-channel and p-channel metal-oxide-semiconductor transistors", *Solid-State Electron.*, vol. 54, no. 4 (April 2010), pp. 461–474.

[52] A Misra and I. Finnie, "On the scribing and subsequent fracturing of silicon semiconductor wafers", *Journal of Materials Science*, vol. 14 (1979), pp. 2567–2574.

[53] G. E. Anner, *Planar Processing Primer*, Van Nostrand Reinhold, New York, 1990, Chapter 2 "Wafers", p. 45, pp. 55–57.

[54] T. Oishi, K. Shiozawa, K. Sugihara, Y. Abe and Y. Tokuda, "Protection of field oxide in trench isolation against contact hole etching to improve alignment tolerance", *Jpn. J. Appl. Phys.*, vol. 37, no. 7B (15 July 1998), pp. L833–L835.

[55] S. Ito, H. Namba, K. Yamaguchi, T. Hirata, K. Ando, S. Koyama, S. Kuroki, N. Ikezawa, T. Suzuki, T. Saitoh, T. Horiuchi, "Mechanical stress effect of etch-stop nitride and its impact on deep submicron transistor design", *IEDM Tech. Dig.*, pp. 247–250 (2000).

[56] H. Liao, P. S. Lee, L. N. L. Goh, H. Liu, J. L. Sudijono, Q. Elgin and C. Sanford, "The impact of etch-stop layer for borderless contacts on deep submicron CMOS device performance — a comparative study", *Thin Solid Films*, vol. 462–463 (2004), pp. 29–33.

[57] J. Tian, B. Zuo, W. Lu, M. Zhou and L. C. Hsia, "Stress modulation of silicon nitride film by initial deposition conditions for transistor carrier mobility enhancement", *Jpn. J. Appl. Phys.*, vol. 49, no. 5, issue 3 (May 2010), pp. 05FB01-1–05FB01-4.

[58] H. S. Yang, R. Malik, S. Narasimha, Y. Li, R. Divakaruni, P. Agnello, S. Allen, A. Antreasyan, J. C. Arnold, K. Bandy, M. Belyansky, A. Bonnoit, G. Bronner, V. Chan, X. Chen, Z. Chen, D. Chidambarrao, A. Chou, W. Clark, S. W. Crowder, B. Engel, H. Harifuchi, S. F. Huang, R. Jagannathan, F. F. Jamin, Y. Kohyama, H. Kuroda, C. W. Lai, H. K. Lee, W.-H., Lee, E. H. Lim, W. Lai, A. Mallikarjunan, K. Matsumoto, A. McKnight, J. Nayak, H. Y. Ng, S. Panda, R. Rengarajan, M. Steigerwalt, S. Subbanna, K. Subramanian, J. Sudijono, G. Sudo, S.-P. Sun, B. Tessier, Y. Toyoshima, P. Tran, R. Wise, R. Wong, I. Y. Yang, C. H. Wann, L. T. Su, M. Horstmann, Th. Feudel, A. Wei, K. Frohberg, G. Burbach, M. Gerhardt, M. Lenski, R. Stephan, K. Wieczorek, M. Schaller, H. Salz, J. Hohage, H.Ruelke, J. Klais, P. Huebler, S. Luning, R. van Bentum, G. Grasshoff, C. Schwan, E. Ehrichs, S. Goad, J. Buller, S. Krishnan, D. Greenlaw, M. Raab, and N. Kepler, "Dual stress liner for high performance sub-45 nm gate length SOI CMOS manufacturing", *IEDM Tech. Dig.*, pp. 1075–1077 (2004).

[59] K. Ishimaru, "45 nm/32 nm CMOS — Challenge and perspective", *Solid-State Electron.*, vol. 52, no. 9 (Sep. 2008), pp. 1266–1273.

[60] K. W. Ang, K. J. Chui, V. Bliznetsov. C. H. Tung, A. Du, N. Balasubramanian, G. Samudra, M. F. Li and Y. C. Yeo, "Lattice strain analysis of transistor structures with silicon-germanium and silicon-carbon source/drain stressors," *Appl. Phys. Lett.*, vol. 86, no. 9 (28 February 2005), pp. 093102-1–093102-3.

[61] E. R. Hsieh and S. S. Chung, "The proximity of the strain induced effect to improve the electron mobility in a silicon-carbon source-drain structure of n-channel metal-oxide-semiconductor field-effect transistors," *Appl. Phys. Lett.*, vol. 96, no. 9 (1 March 2010), pp. 093501-1–095501-3.

[62] P. Yang, W. S. Lau, S. W. Lai, V. L. Lo, L. F. Toh, J. Wang, S. Y. Siah and L. Chan, "Physics behind the overall improvement of n-channel metal-oxide-semiconductor transistors by tensile stress", *J. Appl. Phys.*, vol. 108, no. 3 (1 August 2011), pp. 034506-1–034506-12.

[63] H.-D. Lee and Y.-J. Lee, "Arsenic and phosphorus double ion implanted source/drain junction for 0.25 and sub-0.25-μm MOSFET technology", *IEEE Electron Dev. Lett.*, vol. 20, no. 1 (Jan. 1999), pp. 42–44.

[64] F.-C. Wang and C. Bulucea, "BF_2 and boron double-implanted source/drain junctions for sub-0.25-μm CMOS technology", *IEEE Electron Dev. Lett.*, vol. 21, no. 10 (Oct. 2000), pp. 476–478.

[65] H. K. Kim, S. Y. Ong, E. Quek and S. Chu, "High performance device design through parasitic junction capacitance reduction and junction leakage current suppression beyond 0.1 μm technology", *Jpn. J. Appl. Phys.*, vol. 42, no. 4B (April 2003), pp. 2144–2148.

[66] A. Mineji and S. Shishiguchi, "Ultra shallow junction and super steep halo formation using carbon co-implantation for 65 nm high performance CMOS devices", *International Workshop on Junction Technology (IWJT '06)*, pp. 84–87.

[67] C. F. Tan, J. G. Lee, L. W. Teo, C. Yin, G. Lin, E. Quek and S. Chu, "A carbon co-implantation technique for formation of steep halo for nFET short channel effect improvement and performance boost", *Symp. VLSI Tech.*, (2008), pp. 32–33.
[68] C.-Y. Hu, J. F. Chen, S.-C. Chen, S.-J. Chang, C.-P. Lee and T. H. Lee, "Improved poly gate engineering for 65 nm low power CMOS technology", *J. Electrochem. Soc.*, vol. 157, no. 1 (Jan. 2010), pp. H38–H43.
[69] K. Imai, K. Yamaguchi, N. Kimizuka, H. Onishi, T. Kudo, A. Ono, K. Noda, Y. Goto, H. Fujii, M. Ikeda, K. Kazama, S. Maruyama, T. Kuwata and T. Horiuchi, "A 0.13-μm CMOS technology integrating high-speed and low-power/high-density devices with two different well/channel structures", *IEDM Technical Digest*, (1999), pp. 667–670.
[70] G. G. Shahidi, "SOI technology for the GHz era", *IBM Journal of Research & Development*, vol. 4, no. 2/3 (March/May 2002), pp. 121–131.
[71] R. H. Havemann and J. A. Hutchby, "High performance interconnects: an integration overview", *Proc. IEEE*, vol. 89, no. 5 (May 2001), pp. 586–601.
[72] E. Richard, R. Fox, C. Monget, M. Zaleski, P. Ferreira, A. Guvenilir, P. Brun, E. Oilier, M. Guillermet, M. Mellier, S. Petitdidier, R. Delsol, W. Besling, L. Marinier, G. Imbert, A. Lagha, L. Broussous, M. Rasco, C. Cregut, S. Downey, G. Huang, M. Haond, N. Cave and A. Perera, "Manufacturability and speed performance demonstration of porous ULK (k = 2.5) for a 45 nm CMOS platform", *Symp. VLSI Tech.*, (2007), pp. 178–179.
[73] Y. H. Wang, M. R. Moitreyee, R. Kumar, S. Y. Wu, J. L. Xie, P. Yew, B. Subramanian, L. Shen and K. Y. Zeng, "The mechanical properties of ultra-low-dielectric-constant films", *Thin Solid Films*, vol. 462–463 (Sep. 2004), pp. 227–230.
[74] M. Kodera, Y. Mochizuki, A. Fukuda, H. Hiyama and M. Tsujimura, "Shear stress analyses in chemical mechanical planarization with Cu/porous low-k structure", *Jpn. J. Appl. Phys.*, vol. 46, no. 4B (April 2007), pp. 1974–1980.
[75] F. Liu and M. P. F. Sutcliffe, "Modeling of delamination of ultra low-k material during chemical mechanical polishing", *Tribology Letters*, vol. 25, no. 3 (March 2007), pp. 225–236.
[76] A. S. Brown, "Flat, cheap and under control: Applied Materials' new polishing technology could be the key to the coming generation of microchips", *IEEE Spectrum*, vol. 42, no. 1 (Jan. 2005), pp. 40–45.
[77] L. Economikos, X. Wabg, X. Sakamoto, P. Ong, M. Naujok, R. Knarr, L. Chen, Y. Moon, S. Neo, J. Salfelder, A. Duboust, A. Manens, W. Lu, S. Shruati, F. Liu, S. Tsai and W. Smart, "Integrated electro-chemical mechanical planarization (Ecmp) for future generation device technology", *Proc. IITC (International Interconnect Technology Conference, IEEE)*, (2004), pp. 233–235.
[78] M. Mellier, T. Berger, R. Duru, M. Zaleski, M. C. Luche, M. Rivoire, C. Goldberg, G. Wyborn, K.-L. Chang, Y. Wang, V. Ripoche, S. Tsai, M. Thothadri, W.-Y. Hsu, L. Chen, "Full copper electrochemical

mechanical planarization (Ecmp) as a technology enabler for the 45 and 32 nm nodes", *Proc. IITC (International Interconnect Technology Conference, IEEE)*, (2007), pp. 70–72.

[79] J. Huo, "Electrochemistry in ECMP", in *Microelectronic Applications of Chemical Mechanical Planarization*, edited by Y. Li, Wiley, Hoboken, New Jersey, USA, 2008, pp. 295–318.

[80] L. Economikos, "Planarization technologies involving electrochemical reactions", in *Microelectronic Applications of Chemical Mechanical Planarization*, edited by Y. Li, Wiley, Hoboken, New Jersey, USA, 2008, pp. 319–343.

[81] T. Y. Hoffmann, "Integrating high-k/metal gates: gate first or gate-last", *Solid State Technology*, vol. 53, no. 3 (March 2010), pp. 20–21.

[82] W. S. Lau, P. W. Qian, T. Han, N. P. Sandler, S. T. Che, S. E. Ang, C. H. Tung and T. T. Sheng, "Evidence that N_2O is a stronger oxidizing agent than O_2 for both Ta_2O_5 and bare Si below 1000°C and temperature for minimum low-k interfacial oxide for high-k dielectric on Si", *Microelectronics Reliability*, vol. 47, no. 2–3 (Feb.-Mar. 2007), pp. 429–433.

[83] K. Kita, S. Suzuki, H. Nomura, T. Takahashi, T. Nishimura and A. Toriumi, "Direct evidence of GeO volatilization from GeO_2/Ge and impact of its suppression on GeO_2/Ge metal-insulator-semiconductor characteristics", *Jpn. J. Appl. Phys.*, vol. 47, no. 4 (April 2008), pp. 2349–2353.

[84] H. Kim, P. C. McIntyre, C. O. Chui, K. C. Saraswat and S. Stemmer, "Engineering chemically abrupt high-k metal oxide/silicon interfaces using an oxygen-gettering metal overlayer", *J. Appl. Phys.*, vol. 96, no. 6 (15 September 2004), pp. 3467–3472.

[85] K. Choi, H. Jagannathan, C. Choi, L. Edge, T. Ando, M. Frank, P. Jamison, M. Wang, E. Cartier, S. Zafar, J. Bruley, A. Kerber, B. Linder, A. Callegari, Q. Yang, S. Brown, J. Stathis, J. Iacoponi, V. Paruchuri and V. Narayanan, "Extremely scaled gate-first high-k/metal gate stack with EOT of 0.55 nm using novel interfacial layer scavenging techniques for 22 nm technology node and beyond", *Symp. VLSI Tech.*, (2009), pp. 138–139.

[86] K. Choi, T. Ando, E. Cartier, A. Kerber, V. Paruchuri, J. Iacoponi and V. Narayanan, "The past, present and future of high-k/metal gates", *ECS Trans.*, vol. 53, no. 3 (2013), pp. 17–26.

[87] T. Ando, "Ultimate scaling of high-k gate dielectrics: higher-k or interfacial layer scavenging", *Materials*, vol. 5 (2012), pp. 478–500.

[88] X. Huang, W.-C. Lee, C. Kuo, D. Hisamoto, L. Chang, J. Kedzierski, E. Anderson, H. Takeuchi, Y.-K. Choi, K. Asano, V. Subramanian, T.-J. King, J. Bokor and C. Hu, "Sub 50-nm FinFET: PMOS", *IEDM Technical Digest*, (1999), pp. 67–70.

[89] D. Hisamoto, T. Kaga and E. Takeda, "Impact of the vertical SOI 'Delta' structure on planar device technology", *IEEE Trans. Electron Dev.*, vol. 38, no. 6 (June 1991), pp. 1419–1424.

[90] D. Hisamoto, W.-C. Lee, J. Kedzierski, E. Anderson, H. Takeuchi, K. Asano, T.-J. King, J. Bokor and C. Hu, "A folded-channel MOSFET for deep-sub-tenth micron era", *IEDM Technical Digest*, (1998), pp. 1032–1034.

[91] C. D. Young, K. Akarvardar, M. O. Baykan, K. Matthews, I. Ok, T. Nagai, K.-W. Ang, J. Pater, C. E. Smith, M. M. Hussain, P. Majhi and C. Hobbs, "(110) and (100) sidewall-oriented FinFETs: A performance and reliability investigation", *Solid-State Electron.*, vol. 78, Selected Papers from ISDRS 2011 (December 2012), pp. 2–10.

[92] A. Lubow, S. Ismail-Beigi and T. P. Ma, "Comparison of drive currents in metal-oxide-semiconductor field-effect transistors made of Si, Ge, GaAs, InGaAs and InAs channels," *Appl. Phys. Lett.*, vol. 96, no. 12 (2010), pp. 122105-1–122105-3.

[93] R. Pillarisetty, "Academic and industry research progress in germanium nanodevices", *Nature*, vol. 479, no. 7373 (17 November 2011), pp. 324–328.

[94] S. D. Gupta, J. Mitard, G. Eneman, B. De Jaeger, M. Meuris and M. M. Heyns, "Performance enhancement in Ge pMOSFETs with ⟨100⟩ orientation fabricated with a Si-compatible process flow", *Microelectronic Engineering*, vol. 87, no. 11 (Nov. 2010), pp. 2115–2118.

[95] H. Fang, S. Chuang, K. Takei, H. S. Kim, E. Plis, C.-H. Liu, S. Krishna, Y.-L. Chueh and A. Javey, "Ultrathin-body high-mobility InAsSb-on-insulator field-effect transistors", *IEEE Electron Dev. Lett.*, vol. 33, no. 4 (April 2012), pp. 504–506.

[96] A. Toriumi, C. H. Lee, T. Nishimura, S. K. Wang, K. Kita and K. Nagashio, "Recent progress of Ge technology for a post-Si CMOS", *ECS Trans.*, vol. 35, no. 3 (2011), pp. 443–456.

[97] A. Nainani, T. Irisawa, Z. Yuan, Y. Sun, T. Krishnamohan, M. Reason, B. R. Bennett, J. B. Boos, M. G. Ancona, Y. Nishi and K. C. Saraswat, "Development of high-k dielectric for antimonides and a sub 350°C III-V pMOSFET outperforming germanium", *IEDM Technical Digest*, (2010), pp. 138–141.

[98] A. Nainani, B. R. Bennett, J. B. Boos, M. G. Ancona and K. C. Saraswat, "Enhancing hole mobility in III-V semiconductors", *J. Appl. Phys.*, vol. 111, no. 10 (2012), article number 103706.

[99] R. Pillarisetty, B. Chu-Kung, S. Corcoran, G. Dewey, J. Kavalieros, H. Kennel, R. Kotlyar, V. Le, D. Lionberger, N. Metz, N. Mukherjee, J. Nah, W. Rachmady, M. Radosavljevic, U. Shah, S. Taft, H. Then, N. Zelick and R. Chau, "High mobility strained germanium quantum well field effect transistor as the p-channel device option for low power (Vcc = 0.5 V) III-V CMOS architecture", *IEDM Technical Digest*, (2010), pp. 150–153.

Chapter Four

Low Power CMOS Engineering

4.1 Introduction to Low-power CMOS Engineering

As discussed in Chapter 2, the first generation of microprocessor Intel 4004 was fabricated by PMOS technology. Subsequently, Intel microprocessors were fabricated by NMOS technology. NMOS technology is superior to PMOS technology in terms of switching speed because the electron mobility is bigger than the hole mobility in silicon. NMOS technology suffers from high power consumption, resulting in a lot of heat. Of course, PMOS technology suffers from the same problem. NMOS technology is superseded by CMOS technology, which offers much less power consumption. NMOS technology suffers from high power consumption because while the n-channel MOS transistor serving as the switch is either "ON" or "OFF", the n-channel MOS transistor serving as the load is always "ON". For an integrated circuit based on NMOS technology, approximately 50% of the logic gates are "ON" for both the switch transistor and the load transistor, resulting in large power consumption. CMOS technology uses n-channel MOS transistors and p-channel MOS transistors connected in such a way that in a logic gate when the n-channel MOS transistor is "ON" the corresponding p-channel MOS transistor is "OFF" and vice versa. Thus CMOS technology is supposed to be free of the high power consumption problem of PMOS and NMOS technologies. However, when the number of MOS transistors becomes larger and larger, eventually CMOS integrated circuits also suffer from high power consumption. Table 4.1 shows the power dissipation of some models of microprocessors in the PowerPC series. Currently, there are some microprocessors which consume even more

Table 4.1. Power dissipation of various microprocessors

Model	CMOS technology generation	Clock frequency	Supply voltage for core	Power	Speed to power ratio (MHz/W)
Dual-core PowerPC MPC8641D	90 nm	2 GHz	1.2 V	15–25 W	100
PowerPC 750FX	0.13 μm	900 MHz	1.2 V	3.6 W	250
PowerPC 750CXe	0.18 μm	600 MHz	1.8 V	6 W	100
PowerPC MGT560 (Performance = 56 MIPS)	0.20 μm	56 MHz	2.7 V	0.5 W	112

power than those shown in Table 4.1. The readers should be able to see that newer generations of microprocessors tend to consume much more power than older generations of microprocessors. This is true in general for CMOS digital integrated circuits. To reduce power consumption in CMOS digital integrated circuits, engineers have to pay attention to:

- Circuit-level and system-level engineering (for IDM and design house)
- Process and device engineering (for IDM and foundry)

An IDM (integrated device manufacturer) is a big company which can do the circuit/system level design and also fabricate the designed integrated circuit. An example of IDM is "Intel". An IDM can pay attention to both circuit-level and system-level engineering and also process and device level engineering in order to cut power consumption. For a design house or a fabless company, engineers can pay attention to circuit-level and system-level engineering but not process and device level engineering. For a silicon foundry, engineers can pay attention to process and device level engineering but not circuit-level and system-level engineering. In this book, we will put priority on the process and device engineering part for low-power CMOS engineering. Some circuit-level and system-level power cutting techniques may still be discussed.

4.2 Power Dissipation in CMOS

Rolf Landauer (1927–1999), who was a scientist working in IBM, argued that a computation by itself does not necessarily generate heat but the erasure of a bit of information produces a minimum amount of waste heat.[1] This is known as "Landauer's principle". Indeed, Landauer's principle is very interesting in terms of pure physics and Landauer's 1961 paper has been re-printed several times.[1] However, real computers do not operate under perfect conditions and so generate much more heat than the ideal system considered by Landauer.[2] In 2010, Leland Chang (IBM) and his co-workers published a paper in the Proceedings of IEEE with the title of "Practical strategies for power-efficient computing technologies"; in the beginning, they discussed more conventional principles and then near the end of their article, they discussed Lanaduer's principle and some possible methods to dramatically reduce power consumption.[3] However, this sort of methods may still be far away for a practical engineer. Therefore in this book, for the discussion of heat generation in practical CMOS digital integrated circuits, heat generation does not follow Landauer's principle. For CMOS digital integrated circuits, power dissipation consists of (1) dynamic power dissipation and (2) static power dissipation. For a CMOS inverter,

- Dynamic power dissipation: $C_L V_{DD}^2 f$ (switching at frequency f)
- Static power dissipation: $V_{DD} I_{off}$ (no switching)

For a CMOS inverter,

(1) C_L is the parasitic load capacitance at the output of a CMOS inverter.
(2) $C_L V_{DD}^2/2$ is dissipated in PMOS transistor during charging of C_L.
(3) $C_L V_{DD}^2/2$ is dissipated in NMOS transistor during discharging of C_L.
(4) For each cycle, the energy dissipated is $C_L V_{DD}^2$. Hence, power dissipation is $C_L V_{DD}^2 f$.

4.3 Reduction of Dynamic Power Dissipation

The derivation of the dynamic power dissipation can be found in the paper by Swanson and Meindel.[4] Besides dynamic power dissipation and static power dissipation, there is one additional power dissipation mechanism for a CMOS inverter; this is known as "short circuit power dissipation". There is "short circuit power dissipation" when both the n-channel MOS transistor and the p-channel MOS transistor are turned on briefly during switching for the CMOS inverter. In 1984, Veendrick published a paper to show that the "short circuit power dissipation" can be very much smaller than the "dynamic power dissipation" when the input rise/fall time and the output rise/fall time are short,[5] as shown in Figs. 4.1 and 4.2. Similar discussion can be

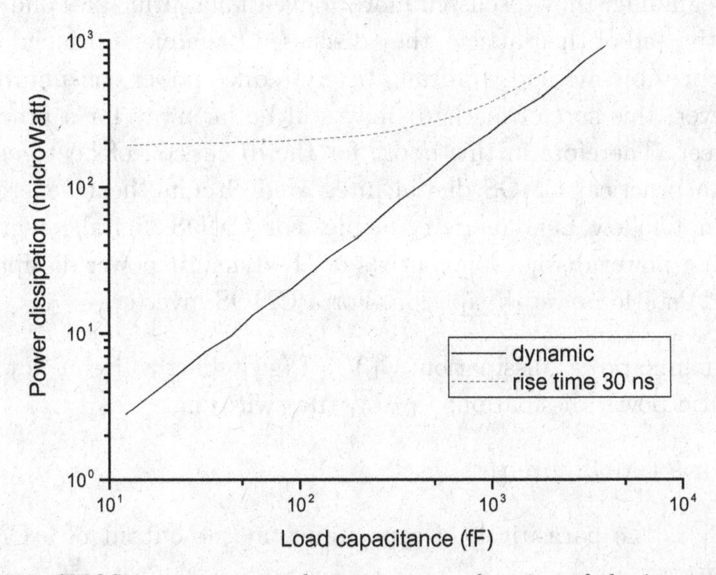

Fig. 4.1 CMOS inverter power dissipation as a function of the inverter load capacitance at the output of a CMOS inverter according to Veendrick 1984.[5] The solid line is the dynamic power dissipation corresponding to 0 input rise time and fall time. The dashed line is the power dissipation corresponding to 30 ns input rise time and fall time. The switching frequency is 10 MHz. (This diagram shows that the short circuit power dissipation is smaller for smaller rise time or fall time; when the rise time or fall time is small enough, the dynamic power dissipation dominates over the short circuit power dissipation.) (Modified from Fig. 6 in H. J. M. Veendrick, "Short-circuit power dissipation estimation for CMOS logic gates", *IEEE J. Solid-State Circuits*, vol. SC-19, no. 4 (Aug. 1984), pp. 468–473.)

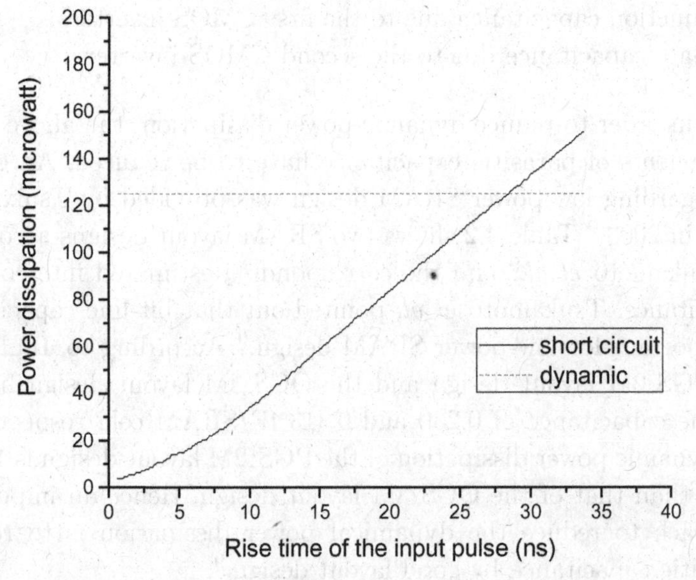

Fig. 4.2 CMOS inverter power dissipation as a function of the rise time of the input pulse according to Veendrick 1984.[5] τ_i is the input rise time and fall time. τ_o is the output rise time and fall time. C_L is the load capacitance at the output of a CMOS inverter. (This diagram shows that the short circuit power dissipation is smaller for smaller rise time or fall time; when the rise time or fall time is small enough, the dynamic power dissipation dominates over the short circuit power dissipation.) (Modified from Fig. 7 in H. J. M. Veendrick, "Short-circuit power dissipation estimation for CMOS logic gates", *IEEE J. Solid-State Circuits*, vol. SC-19, no. 4 (Aug. 1984), pp. 468–473.)

found in the text book by Kang and Leblebici[6] and also in the book "Low-power CMOS circuits" edited by Piguet.[7] Equation (3.67) in the book by van der Meer *et al.*[8] also shows that the "short circuit power dissipation" is proportional to the rise time. Thus when the rise time is reduced, the "short circuit power dissipation" is also reduced.

Assume that two CMOS inverters are cascaded together. The parasitic capacitance C_L at the output of the first CMOS inverter actually has many components. It includes:

(1) The parasitic capacitance due to the interconnect between two CMOS inverters.

(2) Junction capacitance due to the first CMOS inverter.
(3) Gate capacitance due to the second CMOS inverter.

Thus in order to reduce dynamic power dissipation, the above three components of parasitic capacitance have to be reduced. An example regarding low-power SRAM design was provided by Tsukamoto *et al.* in 2003.[9] Table 4.2 shows two SRAM layout designs according to Tsukamoto *et al.*[9] and the corresponding estimated interconnect capacitance. Tsukamoto *et al.* pointed out that bit-line capacitance is important for low-power SRAM design.[9] According to Table 4.2, the PGS_2M layout design and the OGS_3M layout design have a bit-line capacitance of 0.230 and 0.413 fF/SRAM cell, respectively; the dynamic power dissipation of the PGS_2M layout design is 17.9% lower than that of the OGS_3M layout design. Hence an important approach to reduce the dynamical power dissipation is to reduce parasitic capacitance by good layout design.[9]

A method to reduce junction capacitance was suggested by Imai *et al.* in 1999.[10] Basically, Imai *et al.* pointed out that there is an increase in junction capacitance because of the fact that the threshold adjust implant, the anti-punchthrough implant are not masked off in the drain and source area. Imai *et al.* proposed to use two extra masks (one for n-channel and one for p-channel) to mask off those superfluous implants in the drain and source area, resulting in smaller junction capacitance. This decrease in parasitic capacitance

Table 4.2. Estimated interconnect capacitance for each cell (fF/unit memory cell)[9]

SRAM cell layout design	Bit line parasitic capacitance (fF/unit memory cell)	Word line parasitic capacitance (fF/unit memory cell)	Remark
PGS_2M	0.230	0.696	Smaller bit line parasitic capacitance
OGS_3M	0.413	0.334	

Note: Smaller bit line capacitance gives the better low power SRAM.

can speed the transistors and also can somewhat reduce dynamic power dissipation. However, two extra masks are needed, resulting in higher cost.

Lee and Lee reported another approach to reduce junction capacitance in 1999 by using an arsenic implant plus an additional phosphorus implant to form deep drain/source junction, resulting in a more graded n^+ drain/source to p-well junction for n-channel MOS transistors.[11] Similarly, Wang and Bulucea reported another approach to reduce junction capacitance in 2000 by using a BF_2^+ implant plus an additional B^+ implant to form deep drain/source junction, resulting in a more graded p^+ drain/source to n-well junction for p-channel MOS transistors.[12] Of course, these two approaches can be combined together. For example, Kim *et al.* reported their approach to reduce junction capacitance in 2003.[13] For n-channel MOS transistors, Kim *et al.* used an arsenic implant plus an additional phosphorus implant to form deep drain/source junction, resulting in a more graded drain/source to p-well junction; the older approach was to use a single arsenic implant. For p-channel MOS transistors, Kim *et al.* used two different boron implants to form deep drain/source junction, resulting in a more graded drain/source to n-well junction; the older approach was to use a single boron implant. This decrease in parasitic capacitance can speed the transistors and also can somewhat reduce dynamic power dissipation. In fact, any method which can be used to reduce parasitic capacitance can speed up switching speed and cut down dynamic power dissipation. The application of SOI technology can definitely reduce parasitic capacitance and so the dynamic power dissipation of SOI is expected to be lower. As shown in Fig. 4.3, the junction Cj is lower for smaller silicon film thickness.[14,15] When the silicon film thickness becomes larger and larger, the performance of SOI CMOS will approach that of bulk CMOS. Thus, from this figure, the readers can roughly see that the junction capacitance can be reduced by migrating from bulk CMOS to SOI CMOS, resulting in smaller delay time and so better switching speed measured by using the ring oscillator approach. As we have discussed above, the reduction of

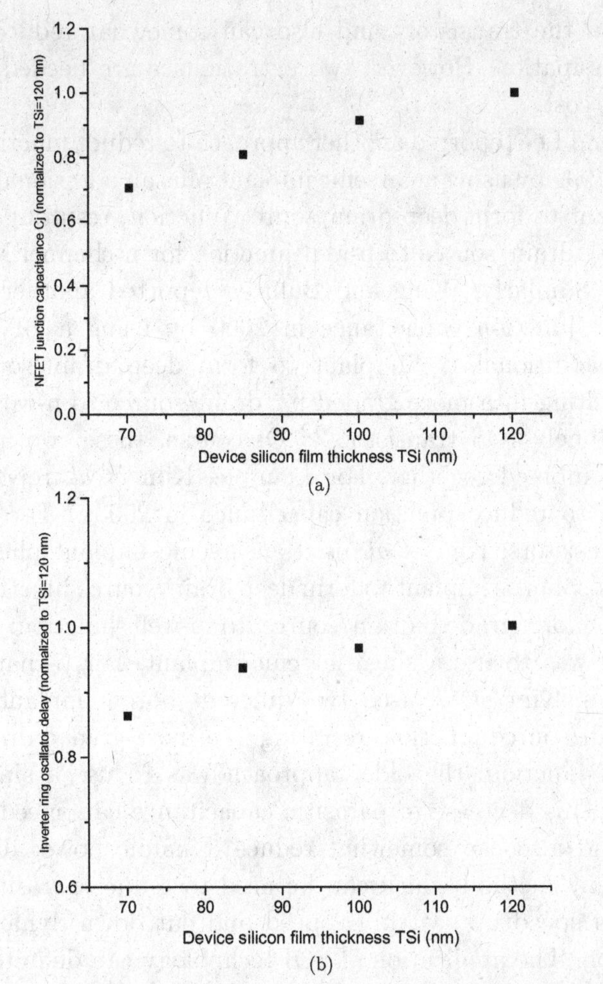

Fig. 4.3 Experimental (a) junction capacitance Cj and (b) ring oscillator performance of SOI according to Sleight *et al.* IEDM 2001.[14] When silicon film thickness becomes larger and larger, the performance of SOI CMOS will approach that of bulk CMOS. Thus, from this figure, the readers can roughly see that the junction capacitance can be reduced by migrating from bulk CMOS to SOI CMOS. (Modified from Fig. 6 in J. W. Sleight, P. R. Varekamp, N. Lustig, J. Adkisson, A. Allen, O. Bula, X. Chen, T. Chou, W. Chu, J. Fitzsimmons, A. Gabor, S. Gates, P. Jamison, M. Khare, L. Lai, J. Lee, S. Narasimha, J. Ellis-Monaghan, K. Peterson, S. Rauch, S. Shukla, P. Smeys, T.-C. Su, J. Quinlan, A. Vayshenker, B. Ward, S. Womack, E. Barth, G. Biery, C. Davis, R. Ferguson, R. Goldblatt, E. Leobandung, J. Welser, I. Yang, and P. Agnello, "A high performance 0.13 μm SOI CMOS technology with a 70 nm silicon film and with a second generation low-k Cu BEOL", *IEDM Technical Digest*, (2001), pp. 245–248.)

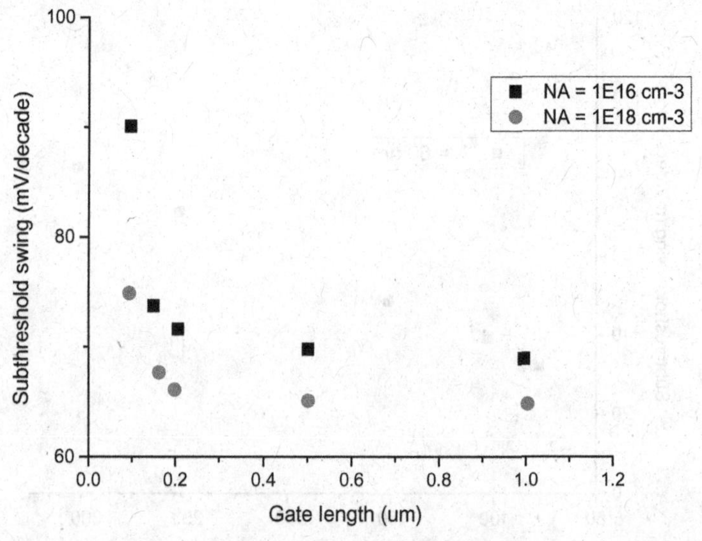

Fig. 4.4 Subthreshold swing (i.e. S-factor) versus gate length for two different channel doping concentrations obtained by a 2-D analytical model developed by Joachim *et al.* 1993.[16] (Modified from Fig. 6 in H.-O. Joachim, Y. Yamaguchi, K. Ishikawa, Y. Inoue and T. Nishimura, "Simulation and two-dimensional analytical modeling of subthreshold slope in ultrathin-film SOI MOSFET's down to 0.1 μm gate length", *IEEE Trans. Electron Dev.*, vol. 40, no. 10 (Oct. 1993), pp. 1812–1817.)

parasitic capacitance will also lead to a reduction of dynamic power dissipation.

As shown in Fig. 4.4, Joachim *et al.* pointed out that the application of SOI technology can lead to a small subthreshold swing of about 60 mV/decade which cannot be easily achieved by conventional CMOS technology.[16] A small subthreshold swing usually lead to smaller static power dissipation. (This will be discussed below.) Of course, this is the case when the gate length is not too small; when the gate length is very small, the subthreshold swing becomes significantly bigger even for SOI technology. Thus SOI technology usually leads to the reduction of both static and dynamic power dissipation. This is good for low power CMOS but SOI technology may also lead to a higher cost. A book was also published by Joachim on SOI.[17]

In addition, as shown in Fig. 4.4, the subthreshold swing becomes high when the gate length becomes very small. This rise in the

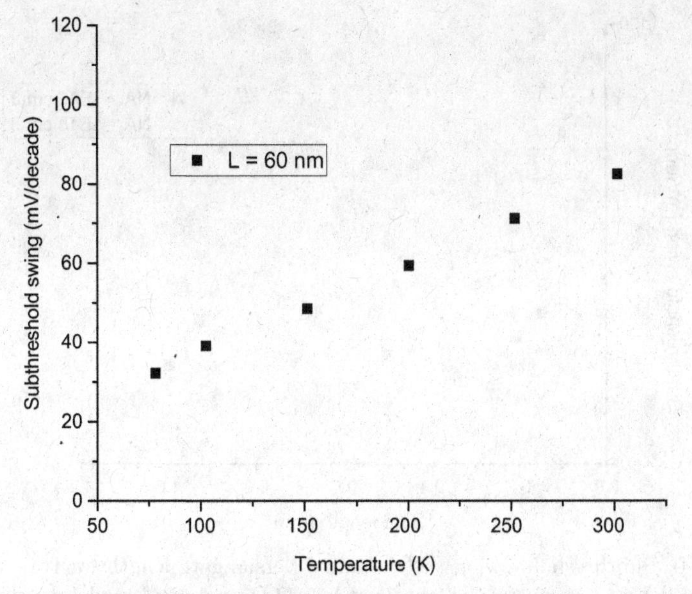

Fig. 4.5 A comparison of the subthreshold swing versus temperature for SOI technology according to Pham-Nguyen *et al.*[19] (Modified from Fig. 5 in L. Pham-Nguyen, C. Fenouillet-Beranger, A. Vandooren, T. Skotnicki, G. Ghibaudo and S. Cristoloveanu, "In situ comparison of Si/high-k and Si/SiO_2 channel properties in SOI MOSFETs", *IEEE Electron Dev. Lett.*, vol. 30, no. 10 (Oct. 2009), pp. 1075–1077.)

subthreshold swing with small gate length looks quite dramatic. This effect has been discussed, for example, by Pouydebasque *et al.*[18]

For CMOS on SOI, cooling down will lead to higher mobility and also lower power dissipation. For example, as shown in Fig. 4.5, the subthreshold swing is further improved by cooling for SOI independent of whether high-k dielectric or conventional silicon dioxide is used for gate dielectric, resulting in further reduction in power dissipation.[19]

4.4 Reduction of Static Power Dissipation

As shown in Fig. 4.6 (modified from Roy *et al.*[20]), there are 6 leakage current mechanisms in deep sub-micron MOS transistors. These leakage currents will lead to static power dissipation. I_1 represents junction leakage current from the drain to substrate (well). I_2 represents

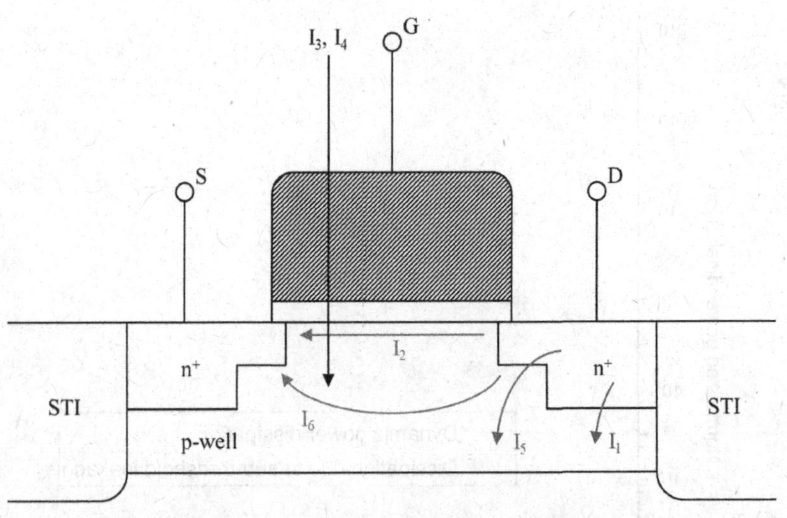

Fig. 4.6 Summary of leakage current mechanisms in deep sub-micron MOS transistors; the author has significantly modified a figure originally from Roy *et al.*[20]: I_1 represents junction leakage current from the drain to substrate (well). I_2 represents sub-threshold leakage current. I_3 is the gate current due to tunneling through the gate dielectric. I_4 is the gate current due to hot carrier injection. I_5 represents gate induced drain leakage (GIDL) current. I_6 represents punchthrough leakage current. (Modified very significantly from Fig. 3 in K. Roy, S. Mukhopadhyay and H. Mahmoodi-Meimand, "Leakage current mechanisms and leakage reduction techniques in deep-submicrometer CMOS circuits", *Proc. IEEE*, vol. 91, no. 2 (Feb. 2003), pp. 305–327.)

sub-threshold leakage current. I_3 is the gate current due to tunneling through the gate dielectric. I_4 is the gate current due to hot carrier injection. I_5 represents gate induced drain leakage (GIDL) current. I_6 represents punchthrough leakage current.

Patel pointed out that for 0.13 μm (130 nm) CMOS technology and the older CMOS technologies, the dynamic power dissipation was the major contributor to the overall CMOS power dissipation while the static power dissipation due to the off current was a minor contributor. Thus in the past, device manufacturers were able to minimize active power consumption by CMOS scaling to newer and smaller process geometries. Smaller geometries allowed MOS devices to run at the same frequency but at a lower voltage, resulting in reduced dynamic power consumption.[21] However, for newer CMOS

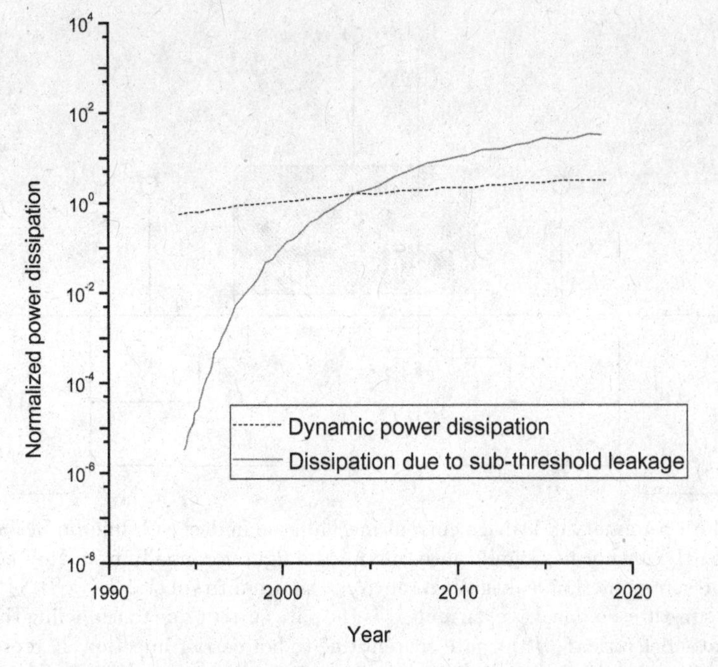

Fig. 4.7 This figure shows that starting from about 2005 the static power dissipation due to subthreshold leakage current became bigger than the dynamic power dissipation according to Kim *et al.* 2003.[22] (Modified from Fig. 1 in N. S. Kim, T. Austin, D. Blaauw, T. Mudge, K. Flautner, J. S. Hu, M. J. Irwin, M. Kandemire and V. Narayanan, "Leakage current: Moore's law meets static power", *Computer*, vol. 36, no. 12 (Dec. 2003), pp. 68–75.)

technology, this may not be the case. As shown in Fig. 4.7, starting from about 2005 the static power dissipation due to subthreshold leakage current became bigger than the dynamic power dissipation according to Kim *et al.* 2003.[22] Thus for newer sub-0.1 μm CMOS technology, the suppression of static power dissipation is very important.

4.4.1 *Reduction of junction leakage current*

The common sense in Si based IC technology is that arsenic implanted n^+/p junction tends to be more leaky than phosphorus implanted n^+/p junction. As discussed in Chapter 1, a heavily doped surface is necessary for forming good Ohmic contact. However, phosphorus diffuses much faster than arsenic such that the surface doping concentration may not be good enough for a good Ohmic contact. It is not too difficult to imagine that the combination of arsenic and phosphorus implantation can be a good idea to make a good n^+/p junction. As discussed above, Lee and Lee reported the application of an arsenic implant plus an additional phosphorus implant to form deep drain/source junction, resulting in a more graded drain/source to p-well junction for n-channel MOS transistors, resulting in lower junction capacitance.[11] They also reported a large improvement in junction leakage, as shown in Fig. 4.8.

Another important source of junction leakage current is related to the salicide process. When there is silicide junction spiking, serious junction leakage can happen. This problem can be solved by optimization of the salicide process and also by making the drain/source junction depth bigger.

Wu *et al.* (TSMC) pointed out that junction leakage current can be improved by optimization of the salicide annealing process (increase of temperature and time), as shown in Fig. 4.9.[23]

Kim *et al.* pointed out that the junction depth can be slightly smaller at STI edge, as shown in Fig. 4.10.[13] Possibility of junction spiking at STI edge should be considered. As discussed above, the use of As/P implant for n-channel MOS transistors and B/B implant for p-channel MOS transistors can help to reduce junction capacitance and thus dynamic power dissipation; the use of As/P implant for n-channel MOS transistors and B/B implant for p-channel MOS transistors also make the junction depth bigger, resulting in less chance of silicide junction spiking.

4.4.2 *Reduction of subthreshold leakage current (part one)*

According to the book "Fundamentals of modern VLSI devices" by Yuan Taur and Tak H. Ning,[24] the subthreshold leakage current is given by the following equation:

$$I_{\mathrm{ds}} = \mu_{\mathrm{sub}} C_{\mathrm{ox}} \frac{W}{L}(m-1)\left(\frac{kT}{q}\right)^2 \exp[q(V_{\mathrm{GS}} - V_{\mathrm{th,sat}})/(mkT)] \times [1 - \exp(-qV_{\mathrm{DS}}/kT)] \quad (4.1)$$

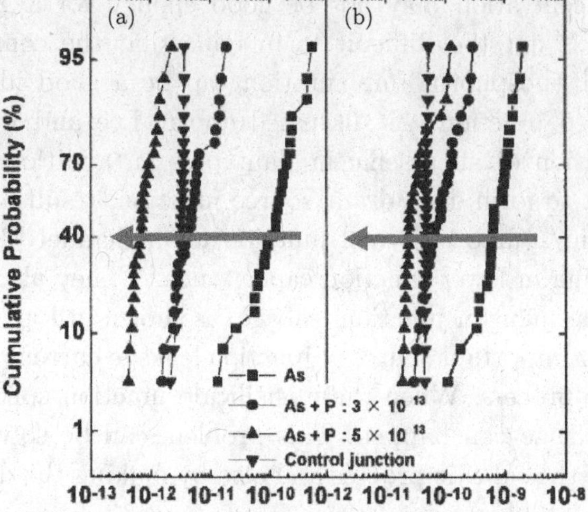

Fig. 4.8 The reverse leakage current characteristics of the As/P implanted n^+ drain/source to p-well junction. (a) Area intensive diode (area: 99856 μm^2, perimeter: 1264 μm) and (b) peripheral intensive diode (area 98000 μm^2, perimeter: 140392 μm) were used for experiments, and the reverse leakage currents were measured at 3 V and 25°C. The reverse leakage current of the proposed As/P implanted junction with a phosphorus dose of 5×10^{13} cm^{-2} was even smaller than that of the non-silicided control junction. The readers should note that titanium salicide technology was used in the paper by Lee and Lee 1999.[11] Subsequently, salicide technology has migrated from titanium salicide to cobalt salicide and then to nickel salicide technology. (Note: The two arrows show that the leakage current is reduced by adding phosphorus for both area intensive and peripheral intensive n+ drain/source to p-well junction diodes.) (Modified from Fig. 2 in H.-D. Lee and Y.-J. Lee, "Arsenic and phosphorus double ion implanted source/drain junction for 0.25- and sub-0.25-μm MOSFET technology", *IEEE Electron Dev. Lett.*, vol. 20, no. 1 (Jan. 1999), pp. 42–44.)

The above treatment from the book by Taur and Ning appears to originate from Swanson and Meindl.[25] In Eq. (4.1), the body-effect coefficient (m) can be expressed as

$$m = 1 + \frac{\sqrt{\varepsilon_0 \varepsilon_{\mathrm{Si}} q N_{\mathrm{ch}} / (4\psi_{\mathrm{B}})}}{C_{\mathrm{ox}}} \tag{4.2}$$

The subthreshold swing S_{ts} is given by the following equation:

$$S_{\mathrm{ts}} = 2.3 \frac{mkT}{q} \tag{4.3}$$

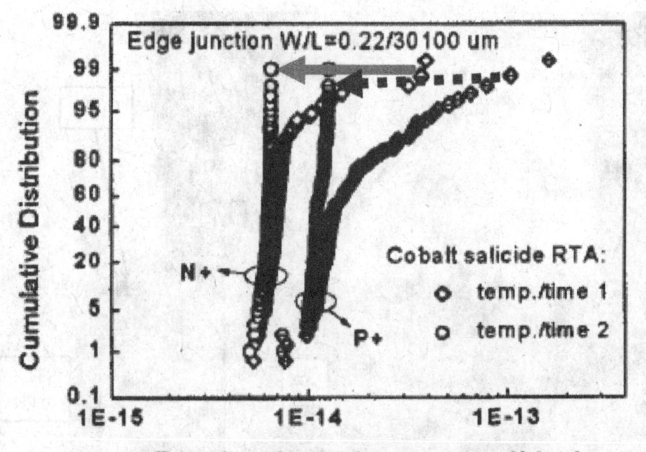

Fig. 4.9 N+ to p-well and P+ to n-well junction leakage for two different $CoSi_2$ RTA conditions according to Wu *et al.* 1999.[23] (Note: The solid arrow shows the reduction in leakage current for N+ to p-well junction diodes due to the change from RTA recipe 1 to RTA recipe 2. The dotted arrow shows the reduction in leakage current for P+ to n-well junction diodes due to the change from RTA recipe 1 to RTA recipe 2.) (Modified from Fig. 11 in C.C. Wu, C.H. Diaz, B.L. Lin, S.Z. Chang, C.C. Wang, J.J. Liaw, C.H. Wang, K.K. Young, K.H. Lee, B.K. Liew and J.Y.C. Sun, "Ultra-low leakage 0.16 mm CMOS for low-standby power applications", *IEDM Technical Digest*, (1999), pp. 671–673.)

The derivation of Eq. (4.3) involves the fact that $\ln 10 = 2.3$.

$$C_{\text{ox}} = \frac{\varepsilon_{\text{o}}\varepsilon_{\text{ox}}}{t_{\text{ox}}} \tag{4.4}$$

According to Fung,[26] $\beta_o = \mu_o \text{C}_{\text{ox}}$ (W/L)

According to Fung,[26] the BSIM subthreshold model was first proposed by B. J. Sheu[27]

The pre-factor was found to be $\beta_o(\text{V}_\text{t})^2\text{e}^{1.8}$ according to Fung,[26] where $\text{Vt} = \text{kT/q}$. Then the subthreshold leakage current is given by the following equation:

$$I_{\text{ds}} = \mu_{\text{sub}} C_{\text{ox}} \frac{W}{L} V t^2 e^{1.8} \exp[q(V_{\text{GS}} - V_{\text{th,sat}})/(mkT)] \times [1 - \exp(-qV_{\text{DS}}/kT)] \tag{4.5}$$

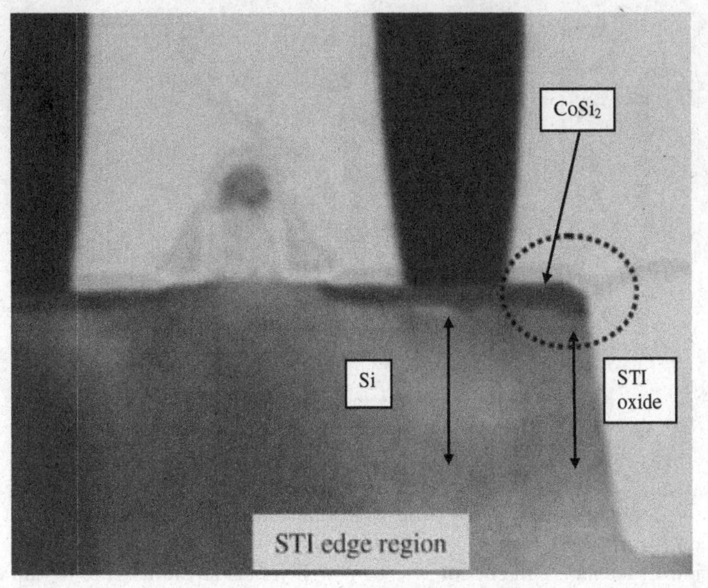

Fig. 4.10 XTEM picture of the STI edge region according to Kim *et al.*[13] The cobalt silicide on the drain (or source) can be seen. However, the XTEM picture does not show the drain (or source) to well junction. Thus the author adds 2 arrows (one longer and one shorter) to show that the junction depth can be smaller at the STI edge. (Modified from Fig. 8 in H. K. Kim, S. Y. Ong, E. Quek and S. Chu, "High performance device design through parasitic junction capacitance reduction and junction leakage current suppression beyond 0.1 μm technology", *Jpn. J. Appl. Phys.*, vol. 42, no. 4B (April 2003), pp. 2144–2148.)

No matter Eq. (4.1) or Eq. (4.5) is used, it can be easily seen that for gate voltage below the threshold voltage, the subthreshold leakage current is reduced by increasing the saturation threshold voltage $V_{th,sat}$ or by reducing "m", which is proportional to the subthreshold swing S as seen by Eq. (4.3). The threshold voltage can be increased by increasing the dose of the threshold adjust implant. The subthreshold swing can be reduced by using thinner gate oxide. The readers should note that Eq. (4.1) or Eq. (4.5) may give an impression that the subthreshold leakage current can be reduced by the decrease of C_{ox} through the increase of the gate oxide thickness. For a constant threshold voltage, thinner gate oxide strongly reduces the subthreshold leakage current through the factor $\exp[q(V_{GS}-V_{th,sat})/(mkT)]$

such that increase of C_{ox} through the decrease of the gate oxide thickness is a minor effect.

As discussed by Pfiester *et al.*[28] and Brews,[29] the threshold voltage is quite frequently chosen at about one quarter of the power supply voltage. To suppress the static power dissipation, one of the simplest approach is to adjust the threshold adjust implant to obtain higher threshold voltage compared to usual design value of threshold voltage, for both n-channel and p-channel MOS transistors. When the magnitude of the threshold voltage is increased, eventually the off current at 0 V gate voltage is dominated by gated induced drain leakage (GIDL) current instead of the subthreshold leakage current.

4.4.3 *Reduction of GIDL*

As shown in Fig. 4.11, the off current at 0 V gate voltage of a p-channel MOS transistor is dominated by GIDL such that further increase in the magnitude of the threshold voltage (for example, by increasing the dose of the threshold adjust implant) cannot reduce the off current. As shown in Fig. 4.11, a stronger poly re-oxidation process increases the thickness of the gate oxide at the poly edge, resulting in lower off current at 0 V gate voltage because of lower GIDL.[23] In addition, a stronger poly re-oxidation process increases the thickness of the gate oxide at the poly edge, resulting in lower off current at 0 V gate voltage because of lower gate to drain/source tunneling current through the gate oxide at the edge.[30] The interested readers should note that a stronger poly re-oxidation process increases the thickness of the gate oxide at the poly edge, resulting in smaller gate to drain/source capacitance and thus somewhat smaller dynamic power dissipation.

In order to explain why poly re-oxidation can help to reduce GIDL, it is necessary to understand the theory of GIDL. In 1987, Chen *et al.*[31] proposed a classical theory of GIDL, which is valid for older generation CMOS technology, as follows.

Let us restrict our discussion to n-channel MOS transistor first; there is a positive voltage V_{DS} between the n-type drain and the n-type source and the voltage between the gate and source V_{GS} is

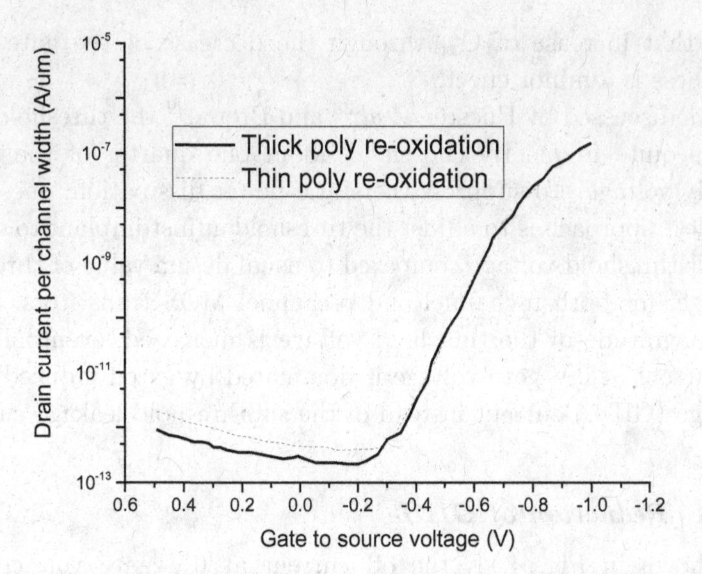

Fig. 4.11 Effect of poly re-oxidation on off current on ULP PMOS transistor according to Wu *et al.*[23] The off current (drain current at zero gate voltage) is smaller for thick poly re-oxidation than for thin poly re-oxidation. (Note 1: W/L = 10/10. $V_{DS} = -2\,V$.) (Note 2: ULP is ultra low power.) (Modified from Fig. 7 in C. C. Wu, C. H. Diaz, B. L. Lin, S. Z. Chang, C. C. Wang, J. J. Liaw, C. H. Wang, K. K. Young, K. H. Lee, B. K. Liew and J. Y. C. Sun, "Ultra-low leakage 0.16 μm CMOS for low-standby power applications", *IEDM Technical Digest*, (1999), pp. 671–673.)

below the threshold voltage such that the MOS transistor is in an "OFF" state. As shown in Fig. 4.12, when the voltage between the drain and gate V_{DG} is large, it is possible that part of the n-type drain region under the gate may be in a state of "deep depletion", resulting in a large surface electric field E_s. The surface electric field is given by

$$E_s = (V_{DG} - 1.2)/(3T_{ox}) \tag{4.6}$$

T_{ox} is the thickness of the gate oxide. In Eq. (4.6), the number 1.2 originates from the fact is the silicon bandgap is about 1.2 eV and the number 3 originates from the fact that the ratio of the dielectric constant of silicon to that of silicon dioxide is about 3.

$$I_{GIDL} = A_{GIDL} E_s \exp(-B_{GIDL}/E_s) \tag{4.7}$$

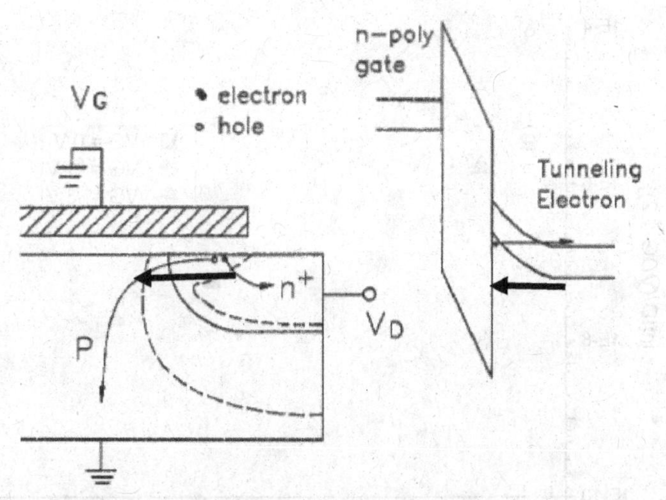

Fig. 4.12 A deep depletion region is formed in the gate-to-drain overlap region. The energy-band diagram illustrates the band-to-band tunneling process and the flow of carriers. Valence-band electrons tunnel into the conduction band and collected at the drain. The holes flow to the p-type substrate or p-well. (Chen *et al.* 1987.[31]) (Note: The two thick arrows show the direction of the electric field in the depletion region of the n+-p junction.) (Modified from Fig. 3 in J. Chen, T.Y. Chan, I.C. Chen, P.K. Ko and C. Hu, "Subbreakdown drain leakage current in MOSFET", *IEEE Electron Dev. Lett.*, vol. EDL- 8, no. 11 (Nov. 1987), pp. 515–517.)

In Eq. (4.7), A GIDL is a pre-exponential constant and B_{GIDL} is given by the following equation:

$$B_{GIDL} = [\pi(m^*)^{1/2}(E_G)^{3/2}]/[(2)^{3/2}q\hbar] \tag{4.8}$$

According to Chen *et al.*,[31] $B_{GIDL} = 21.3\,MV/cm$ with $m^* = 0.2\,m_o$.

According to Eq. (4.6), E_s is poportional to $V_{DG}-1.2$. Figure 4.13 shows that experimental $\ln(I_{GIDL}/(V_{DG} - 1.2))$ plotted against $1/(V_{DG} - 1.2)$ can be easily fitted to a straight line in agreement with Eq. (4.7).

Inspection of Eq. (4.7) shows that I_{GIDL} increases strongly with the increase in E_s. Thus GIDL can be reduced by a reduction of E_s. According to Eq. (4.6), E_s can be reduced by increasing the gate dielectric thickness T_{ox}. However, a larger T_{ox} will degrade the on

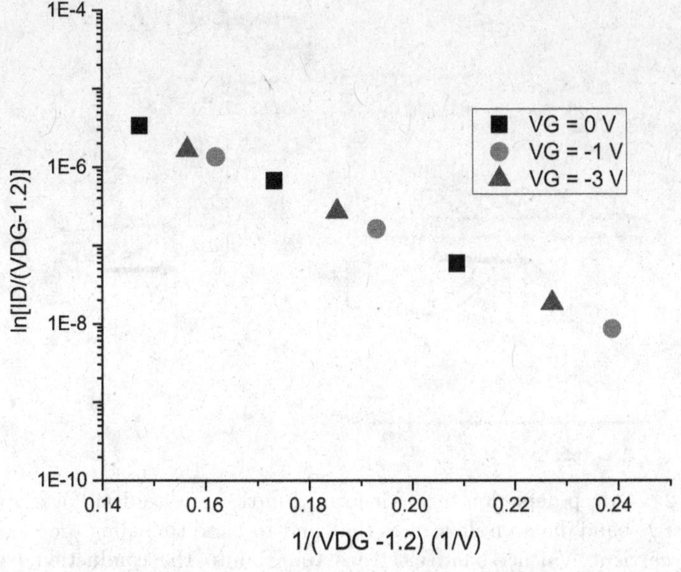

Fig. 4.13 A plot of experimentally measured $\ln(I_{GIDL}/(V_{DG} - 1.2))$ against $1/(V_{DG} - 1.2)$, where V_{DG} is the voltage difference between the drain and the gate. All measured data with the gate oxide thickness fall on a straight line, in agreement with the band-to-band tunneling model. (Chen *et al.* 1987.[31]) The readers should note that the data shown in this figure are from old MOS technology in the 1980's. At that time, the power supply voltage was much larger than 1.2 V. Nowadays, the power supply voltage can be smaller than 1.2 V and the model according to Chen *et al.* 1987 is no longer valid for state-of-the-art MOS technology. (Modified from Fig. 4 in J. Chen, T. Y. Chan, I. C. Chen, P. K. Ko and C. Hu, "Subbreakdown drain leakage current in MOSFET", *IEEE Electron Dev. Lett.*, vol. EDL-8, no. 11 (Nov. 1987), pp. 515–517.)

current. However, T_{ox} can be locally increased by gate re-oxidation at the gate edge, resulting in lower E_s and thus GIDL. Thus a local increase of gate oxide thickness at the gate edge can be used to suppress GIDL without causing serious on current degradation.

The readers should note that the experimental data shown in Fig. 4.13 came from old MOS technology in the 1980's. The power supply voltage for old CMOS technology in the 1980's was 5 V. The power supply voltage for 0.12 μm CMOS technology has already been scaled down to 1.2 V. Further scaling of CMOS technology reduces the power supply voltage to 1.0 V or even smaller. It can be easily imagined that $(V_{DG} - 1.2)$ can become negative for n-channel MOS

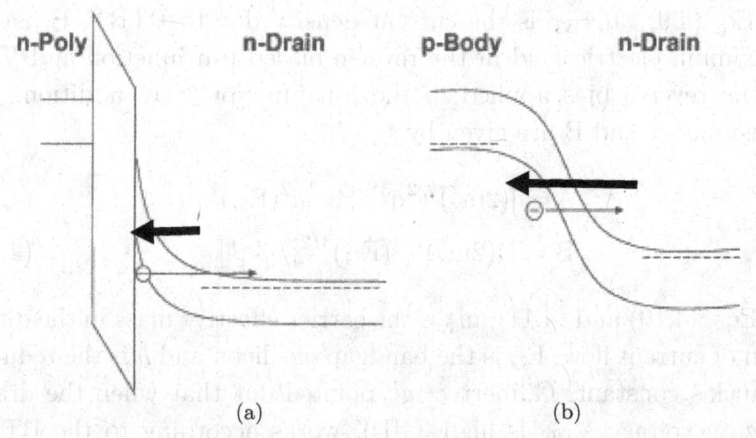

Fig. 4.14 Energy band diagrams of n-channel MOS transistors for the illustration of the physical mechanisms for GIDL according to Yuan *et al.*[32] (a) In the gate-to-drain overlap region, electron tunneling from valence band to conduction band happens when the gate voltage is low enough (or in other words, negative enough) to cause a band bending larger than the silicon bandgap. (Note: This has been the more classical situation for GIDL.) (b) In the p-type doped channel to n-type doped drain junction near the interface of the gate dielectric and silicon, electrons can tunnel from the valence band in the channel to the conduction band in the drain. (Note: This may be the situation for GIDL in newer CMOS technology.) (Note: The two thick arrows show the direction of the electric field in the depletion region of the n+-p junction.) (Modified from Fig. 2 in X. Yuan, J.-E. Park, J. Wang, E. Zhao, D.C. Ahlgren, T. Hook, J. Yuan, V.W.C. Chan, H. Shang, C.-H. Liang, R. Lindsay, S. Park and H. Choo, "Gate-induced-drain-leakage current in 45-nm CMOS technology," *IEEE Transactions on Device and Materials Reliability*, vol. 8, no.3 (Sept. 2008), pp. 501–508.)

transistors because $V_{DG} < 1.2$ and the classical GIDL theory discussed above may not be valid. In fact, Yuan *et al.* (IBM) pointed out that the situation for newer CMOS technology can be somewhat different from the classical GIDL model.[32] As shown in Fig. 4.14, Fig. 4.14(a) corresponds to the old situation while Fig. 4.14(b) corresponds to the situation in newer CMOS technology.

The modeling of GIDL can be complicated in newer CMOS technology. However, GIDL can be understood in a relatively simple manner as BTBT (band-to-band tunneling) of the reverse biased n^+ drain to p-well junction just under the gate dielectric for an n-channel transistor. The equation of BTBT is given as follows[32]:

$$J_{BTBT} = AE_jV_{app}\exp(-B/E_j) \qquad (4.9)$$

In Eq. (4.9), J_{BTBT} is the current density due to BTBT, E_j is the maximum electric field in the reverse biased p-n junction and V_{app} is the reverse bias applied to the p-n junction.[32] In addition, the constants A and B are given by

$$A = [(2m_x)^{1/2}q^3]/[4\pi^3\hbar^2(E_G)^{1/2}] \quad (4.10)$$

$$B = [4(2m_x)^{1/2}(E_G)^{3/2}]/[3q\hbar] \quad (4.11)$$

In Eqs. (4.10) and (4.11), m_x is the carrier effective mass in the direction of current flow, E_G is the bandgap of silicon and $\hbar$ is the reduced Planck's constant. Gilibert *et al.* pointed out that when the drain-to-gate voltage V_{DG} is high, GIDL works according to the BTBT mechanism; however, when V_{DG} is low, GIDL works according to the trap-assisted-tunneling (TAT) mechanism.[33] (Note: In the newer CMOS technology, the channel-to-drain p-n junction responsible for GIDL behaves like a p-n junction which is heavily doped on both sides. The readers should take note that the theory of a p-n junction which is heavily doped on both sides is different from the theory of a p-n junction which is lightly doped on at least one side. For the former, reverse leakage current may be due to tunneling current while for the latter, reverse leakage current may be due to generation current.)

Besides silicon-based CMOS, GIDL is expected to be present in germanium-based CMOS. As shown in Fig. 4.15, GIDL (or BTBT) can be seen in germanium-based MOS transistor.[34] In addition, it can be seen that GIDL (or BTBT) can be reduced by using GeOI (germanium on insulator). As shown in Fig. 4.16, it can also be reduced by "transistor stacking".

It is known that mechanical stress can enhance GIDL. For example, Cheng *et al.* reported that mechanical stress from STI (shallow trench isolation) can enhance GIDL.[35]

Lau *et al.* made an in-depth analysis of the effect of mechanical stress on GIDL for mechanical stress sensor application.[36] Figure 4.17 shows that GIDL changes very strongly with mechanical stress while the on current or threshold voltage do not change strongly with mechanical stress.

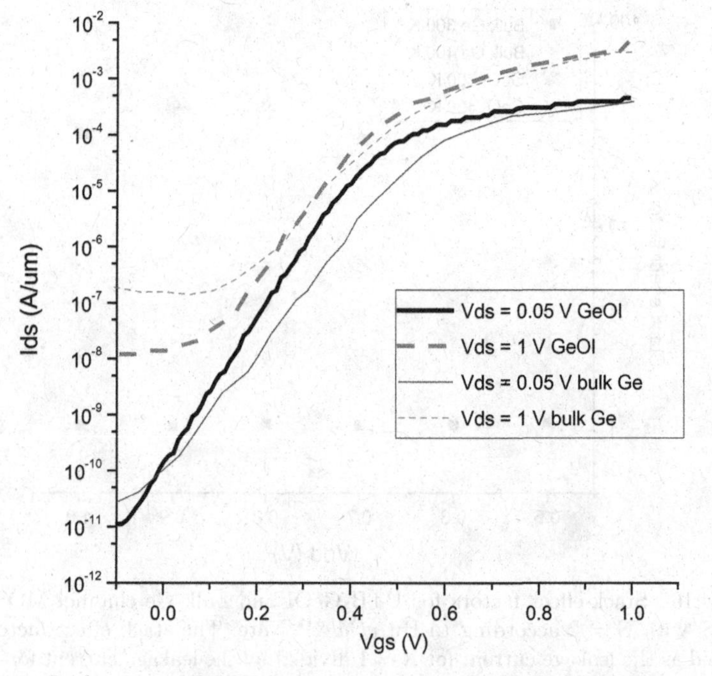

Fig. 4.15 Reduction of BTBT leakage current by UTB GeOI device compared to bulk Ge device according to Hu *et al.*[34] This figure shows the I_{ds}–V_{gs} characteristics at V_{ds} = 0.05 and 1 V for UTB GeOI device and bulk Ge device. They were designed with equal on current at $V_{gs} = V_{ds}$ = 1 V. The saturation threshold voltage Vth-sat is 0.31 V for GeOI and 0.312 V for bulk Ge. However, DIBL is 58 mV/V for GeOI and 118 mV/V for bulk Ge. Thus DIBL is much better for GeOI compared to bulk Ge. (Modified from Fig. 1 in V.P-H. Hu, M.-L. Fan, P. Su and C.-T. Chuang, "Band-to-band-tunneling leakage suppression for ultra-thin-body GeOI MOSFETs using transistor stacking", *IEEE Electron Dev. Lett.*, vol. 33, no. 2 (Feb. 2012), pp. 197–199.)

GIDL can be enhanced by (a) bandgap reduction or by (b) effective electron mass reduction. As shown in Fig. 4.18, the silicon bandgap actually has a "sharp" point at zero mechanical strain such that the bandgap can be reduced by either tensile or compressive stress. However, the effective electron mass plotted against mechanical stress is a smooth curve such that it can decrease with tensile stress while it can increase with compressive stress.

In an actual manufacturing environment, the parasitic mechanical stress in the MOS transistor may have a statistical spread such

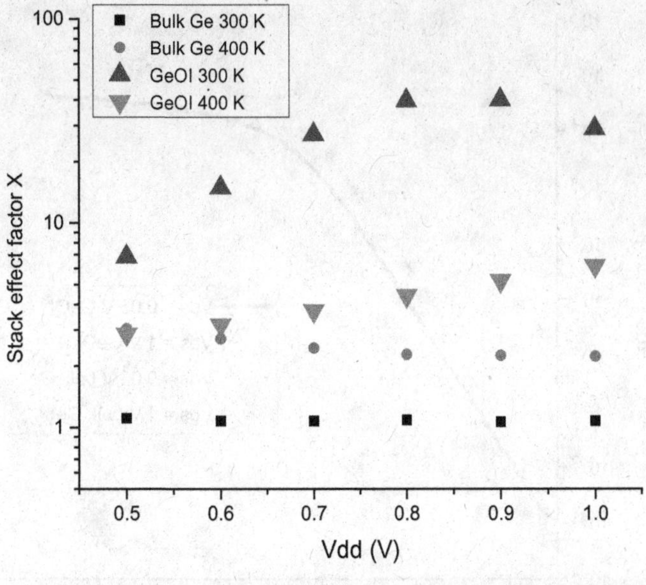

Fig. 4.16 Stack-effect factors for UTB GeOI and bulk Ge channel MOS transistors with N = 2 according to Hu *et al.*[34] (Note: The stack effect factor X is defined as the leakage current for N = 1 divided by the leakage current for N = 2, where N is the number of stacked devices.) At room temperature, the stack effect factor X is approximately equal to 1 and so transistor stacking does not help for the bulk Ge channel devices. The UTB GeOI devices show much larger stack-effect factor X than that for the bulk Ge channel devices at room temperature. At 400 K, the stack effect factor X is not too bad even for the bulk Ge channel devices. (Modified from Fig. 2 in V.P-H. Hu, M.-L. Fan, P. Su and C.-T. Chuang, "Band-to-band-tunneling leakage suppression for ultra-thin-body GeOI MOSFETs using transistor stacking", *IEEE Electron Dev. Lett.*, vol. 33, no. 2 (Feb. 2012), pp. 197–199.)

that it may be "tensile" or "compressive", as shown in Fig. 4.19. As shown in Fig. 4.19, a tensile mechanical stress bias can be used such that an extra tensile stress will decrease both the silicon bandgap and the electron effective mass, resulting in best sensitivity.

4.4.4 *Reduction of subthreshold leakage current (part two)*

When the off current at 0 V gate voltage is dominated by the subthreshold leakage current, there are two main approaches to

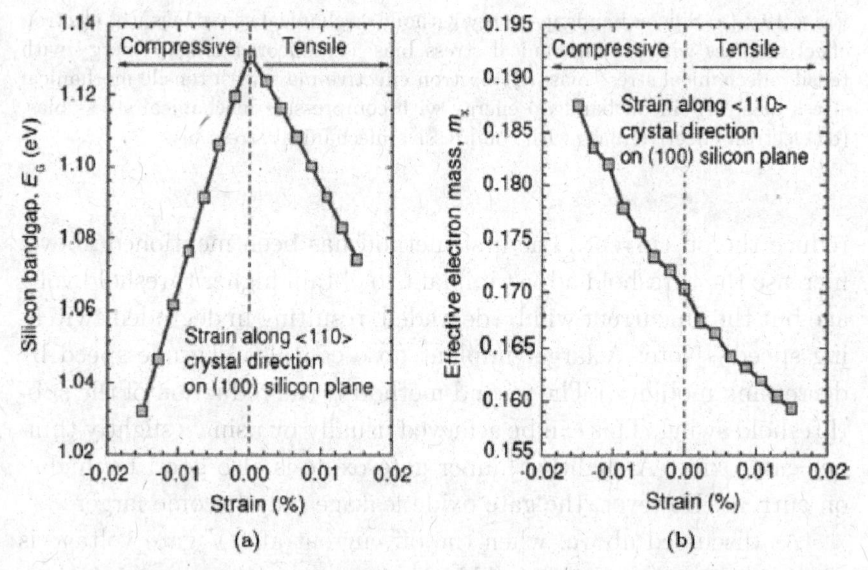

Fig. 4.17 I_{ds} versus V_{GS} characteristics of an n-channel MOS transistor (gate length = 60 nm, gate width = 2 μm) measured from the drain side: (a) Control device. (b) Tensile stressed device. (c) Comparison between control and tensile stressed devices that are in saturation mode.

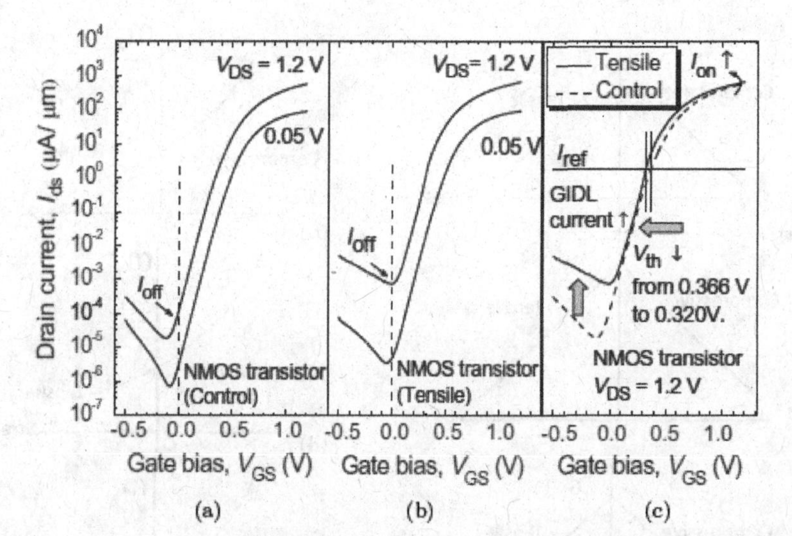

Fig. 4.18 The effects of uniaxial stress on (a) the silicon bandgap and (b) the electron effective mass along the channel direction using tight-binding band structure calculation. The stress was applied along the <110> direction on (100) Si plane in the simulation.

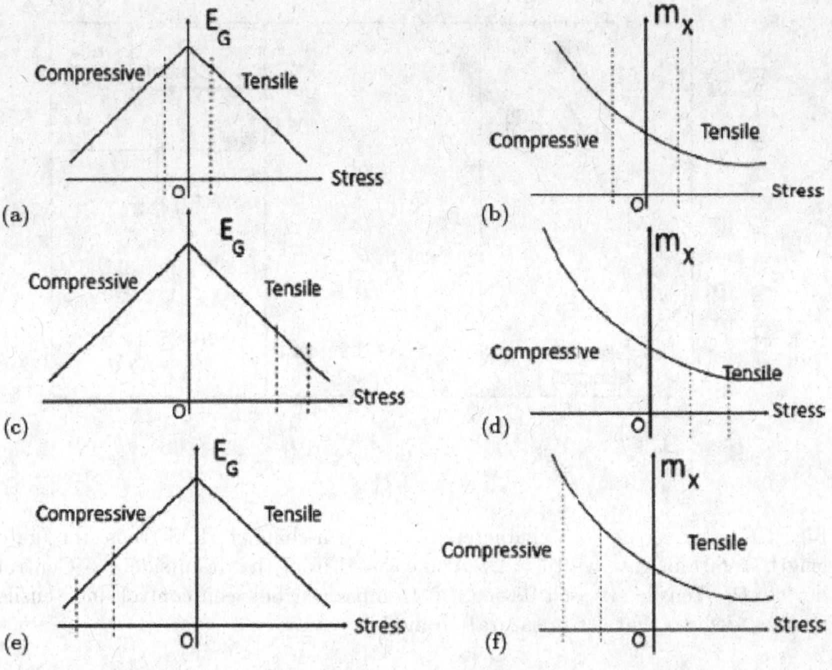

Fig. 4.19 (a) Silicon bandgap energy without mechanical stress bias; (b) electron effective mass without mechanical stress bias; (c) silicon bandgap energy with tensile mechanical stress bias; (d) electron effective mass with tensile mechanical stress bias; (e) silicon bandgap energy with compressive mechanical stress bias; (d) electron effective mass with compressive mechanical stress bias.

reduce the off current. The first method has been mentioned above: increase the threshold adjust implant to obtain higher threshold voltage but the on current will be degraded, resulting in degraded switching speed. (Note: A larger implant dose can also degrade speed by decreasing mobility.) The second method is the reduction of the subthreshold swing. This can be achieved usually by using a slightly thinner gate oxide. A slightly thinner gate oxide is also good for higher on current. However, the gate oxide leakage may become larger.

As discussed above, when the off current at 0 V gate voltage is dominated by the subthreshold leakage current, there are two important parameters: 1) threshold voltage and 2) subthreshold swing. In addition, there is an extra problem due to statistical spread in the gate length or effective channel length. Short channel effect (SCE) is the reduction of the threshold voltage when the gate length or effective channel length becomes smaller. Reverse short channel effect (SCE) is the increase of the threshold voltage when the gate length or effective channel length becomes smaller. SCE and RSCE can happen for the same silicon wafer such that the threshold voltage increases for a certain range of gate lengths but decreases for another range of gate lengths; that is, the threshold voltage can be a complicated function of gate length. The "target" gate length may happen to be in a range of gate lengths suffering from SCE. The actual gate length may be a statistical distribution with the mean at the "target" gate length. The MOS transistors with gate length at the lower limit of the statistical distribution will have significantly higher off current than those MOS transistors with longer gate lengths such that the off current of the whole chip is controlled by those MOS transistors with gate length at the lower limit of the statistical distribution. Thus the static power dissipation of the whole chip can be dominated by those MOS transistors with gate length at the lower limit of the statistical distribution. The solution to this problem is to design the process such that the threshold voltage is not sensitive to minor statistical variation in the gate length at the "target" gate length. One method proposed by Thompson *et al.* in 2005 is to use the "halo" implant to "flatten" the transistor off-state leakage versus gate length characteristics, as shown in Fig. 4.20.[37]

This method appears to be very good but there is at least one known disadvantage of this method: the use of halo implant can degrade the analog performance of the MOS transistor.[38–41] In analog or mixed-signal CMOS integrated circuits, the halo implant may have to be blocked for the analog areas. The degradation of MOS transistor due to the halo implant will be discussed in another chapter of this book.

However, Thompson *et al.*[37] only very briefly mentioned their approach and did not give real data in their paper. The author (W. S. Lau) made a study by himself.[42] The real situation appears to be much more complicated than that represented by Thompson *et al.*,[37] as shown in Fig. 4.21 for n-channel MOS transistors and as shown in Fig. 4.22 for p-channel MOS transistors. The major difference between the author's work[42] and Thompson *et al.*[37] is that

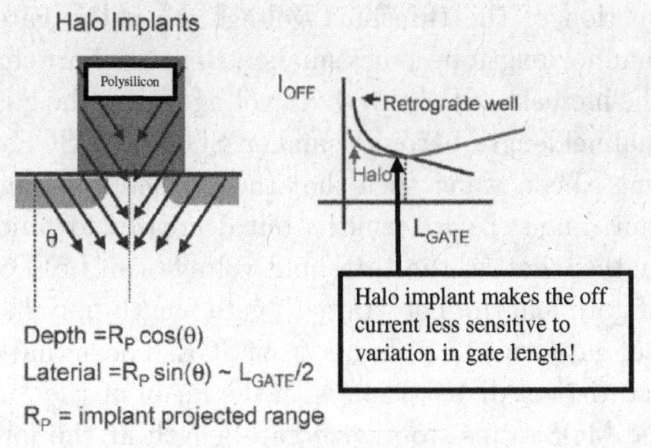

Fig. 4.20 Halo implants used to "flatten" transistor off-state leakage versus gate length. Used to reduce leakage in subdesign rule structures according to Thompson *et al.* 2005.[37] (Modified from Fig. 18 in S.E. Thompson, R.S. Chau, T. Ghani, K. Mistry, "In search of "Forever," continued transistor scaling one new material at a time," *IEEE Trans. Semi. Manuf.*, vol. 18, no. 1 (Feb. 2005), pp. 26–36.)

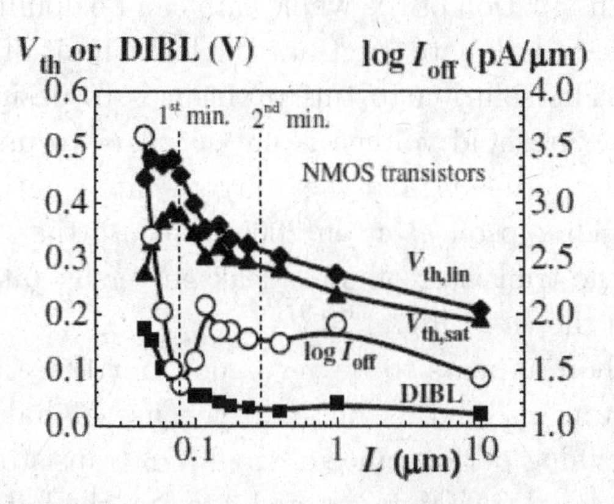

Fig. 4.21 The linear threshold voltage and the saturation threshold voltage as a function of the mask gate length for n-channel MOS transistors.[42] (Reprinted with permission from W. S. Lau, P. Yang, E. T. L. Ng, Z. W. Chian, V. Ho, S. Y. Siah and L. Chan, "Region of nearly constant off current versus gate length characteristics for sub-0.1 μm low power CMOS technology", *Proc. IEEE EDSSC* 2008, pp. 231–234. Copyright 2008 IEEE.)

it is possible to have more than one minimum in the MOS transistor off-state leakage current versus gate length characteristics.

It can be easily imagined that when the gate length becomes very short in look-ahead MOS transistor test structures, the halo implant

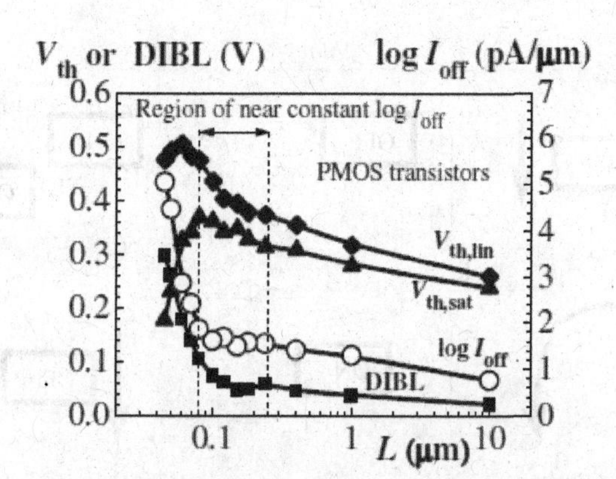

Fig. 4.22 The linear threshold voltage and the saturation threshold voltage as a function of the mask gate length for p-channel MOS transistors.[42] (Reprinted with permission from W. S. Lau, P. Yang, E. T. L. Ng, Z. W. Chian, V. Ho, S. Y. Siah and L. Chan, "Region of nearly constant off current versus gate length characteristics for sub-0.1 μm low power CMOS technology", *Proc. IEEE EDSSC* 2008, pp. 231–234. Copyright 2008 IEEE.)

originally intended for the middle of the channel region may cross over to the drain or source region. In 2010, Lau *et al.* (the author's research group) pointed out that this is indeed possible.[43]

4.4.5 *Reduction of gate leakage current by high-k dielectric technology*

As discussed before, ideal CMOS integrated circuits has zero power dissipation. Power dissipation due to subthreshold leakage current is important in practical non-ideal CMOS integrated circuits. As shown in Fig. 4.23, it is possible to have leakage current passing from the positive supply to the negative supply through "gate leakage" in a 6T SRAM cell according to Geens and Dehaene;[44] for example, the leakage current can go through the source and the gate of the p-channel MOS transistor (right top in Fig. 4.23), which is "ON", and then the gate and the source of the n-channel MOS transistor (left bottom in Fig. 4.23), which is "ON", resulting in extra power dissipation. This is also true for CMOS technology in general. The solution of this problem is to use a dielectric material with a dielectric constant higher than that of silicon dioxide, which is known as high-k dielectric. As shown in Fig. 4.24, using high-k technology instead of

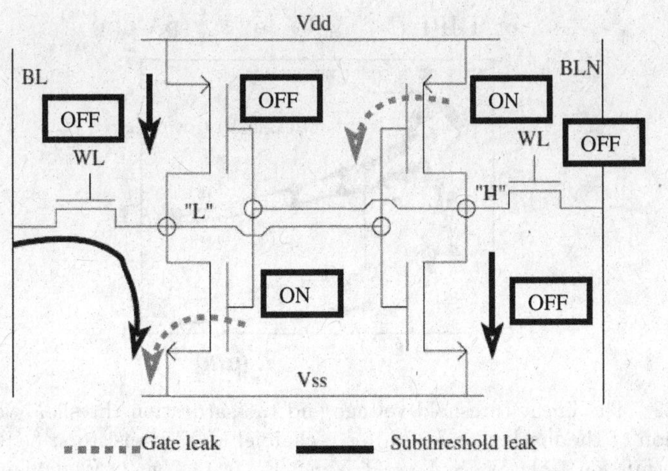

Fig. 4.23 Leakage currents in a 6T SRAM cell according to Geens and Dehaene.[44] (Note: 6T stands for 6 transistors.) (Modified from Fig. 1 in P. Geens and W. Dehaene, "A small granular controlled leakage reduction system for SRAMs", *Solid-State Electron.*, vol. 49, no. 11 (Nov. 2005), pp. 1776–1782.)

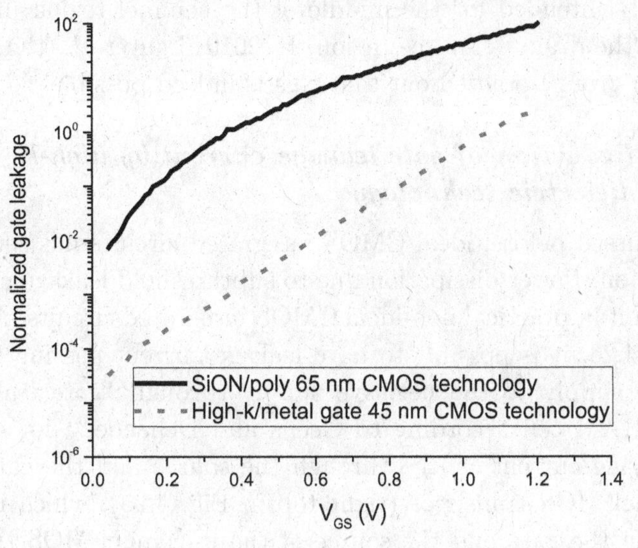

Fig. 4.24 Normalized gate leakage versus gate voltage showing high-k technology can very greatly decrease gate leakage current according to Auth.[45] (Note: The above figure is for n-channel MOS transistors. Qualitatively, the situation is the same for p-channel MOS transistors.) (Modified from Fig. 10 in C. Auth, "45 nm high-k metal gate strained-enhanced CMOS transistors", *Proc. IEEE 2008 Custom Integrated Circuits Conference (CICC)*, pp. 379–386.)

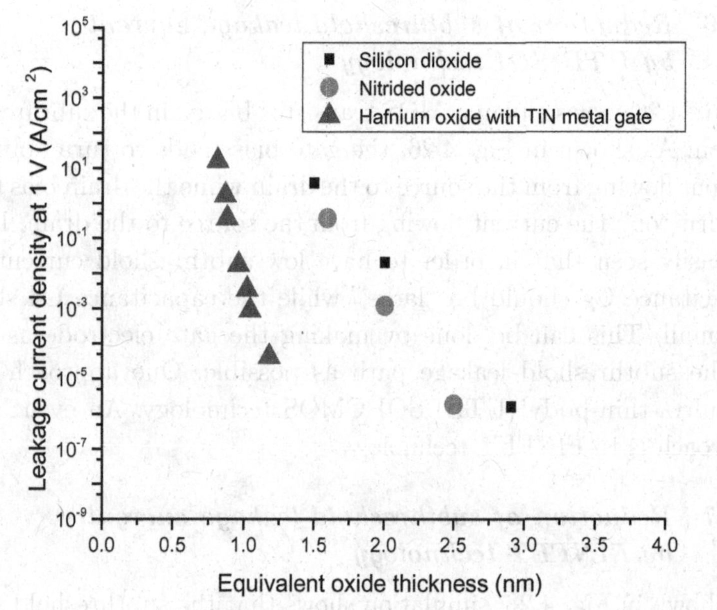

Fig. 4.25 Leakage current density versus EOT (equivalent oxide thickness) for a high-k dielectric material HfO_2 with TiN metal gate compared to SiO_2 and nitrided SiO_2 according to a review paper by Robertson and Wallace.[46] (Modified from Fig. 9 in J. Robertson and R. M. Wallace, "High-k materials and metal gates for CMOS applications", *Materials Science and Engineering R*, vol. 88 (2015), pp. 1–41.)

more conventional SiON gate dielectric can very greatly decrease gate leakage according to Auth.[45] (The interested readers should note that metal gate technology is quite frequently used together with high-k technology to improve CMOS performance.) As shown in Fig. 4.25, the gate leakage current can be significantly reduced by various high-k dielectric materials (for example, HfO_2, ZrO_2, Al_2O_3 and La_2O_3) compared to SiO_2 or SiON according to a review paper by Robertson and Wallace.[46]

In IEDM 2006, Gurfinkel *et al.*[47] pointed out that bulk traps in high-k dielectric can contribute to GIDL in addition to bulk traps in bulk silicon. This was followed up by a journal paper in 2009.[48] In 2008, Liao *et al.* also shared the similar idea that bulk traps in high-k dielectric can contribute to GIDL in addition to bulk traps in bulk silicon.[49]

4.4.6 *Reduction of subthreshold leakage current by UTB SOI technology*

Figure 4.26 shows a planar MOS transistor biased in the subthreshold region. As shown in Fig. 4.26, the gate bias tends to turn "off" the current flowing from the source to the drain while the drain bias tends to turn "on" the current flowing from the source to the drain. It can be easily seen that in order to have low subthreshold current, the capacitance Cg should be "large" while the capacitance Cd should be small. This can be done by making the gate electrode as close to the subthreshold leakage path as possible. One approach is to use ultra-thin-body (UTB) SOI CMOS technology. An even better approach is to FINFET technology.

4.4.7 *Reduction of subthreshold leakage current by FINFET technology*

As shown in Fig. 4.28, simulation shows that the subthreshold swing S of FINFETs is much better than single-gate transistors according

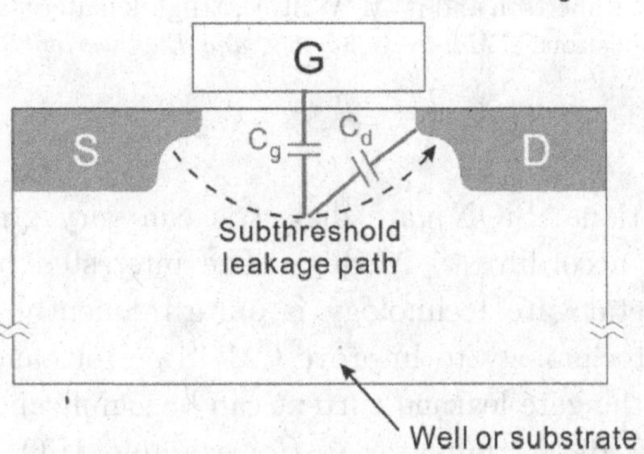

Fig. 4.26 This figure shows a planar MOS transistor biased in subthreshold region. The gate bias tends to turn "off" the current flowing from the source to the drain while the drain bias tends to turn "on" the current flowing from the source to the drain. It can be easily seen that in order to have low subthreshold current, the capacitance Cg should be "large" while the capacitance Cd should be small. This can be done by making the gate electrode as close to the subthreshold leakage path as possible. Experimental support of the above theory can be found from the work of Choi *et al.*[50] as follows. As shown in Fig. 4.27, it can be easily seen that the subthreshold swing becomes smaller with the decrease of the silicon body thickness T_{Si}, resulting in significantly lower off current.

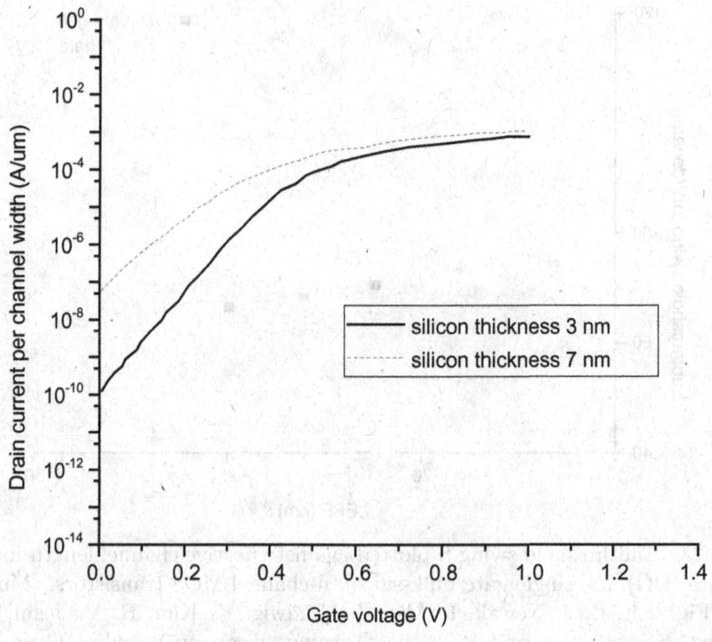

Fig. 4.27 Impact of silicon body thickness T_{Si} on the $I_{ds} - V_{gs}$ characteristics of UTB SOI device ($T_{ox} = 1.5\,nm$, $N_{sub} = 1 \times 10^{15}\,cm^{-3}$ and $V_{ds} = 1\,V$) according to Choi *et al.*[50] (Note: BOX stands for "buried oxide".) (Modified from Fig. 1 in Y.-K. Choi, K. Asano, N. Lindert, V. Subramanian, T.-J. King, J. Bokor and C. Hu, "Ultrathin-body SOI MOSFET for deep-sub-tenth micron era", *IEEE Electron Dev. Lett.*, vol. 21, no. 5 (May 2000), pp. 254–255.)

to Nowak *et al.*[51] As discussed before, a better S is important for low power CMOS applications. Experimentally, Table 4.3 shows that FINFET has very good subthreshold swing S for both n-channel and p-channel transistors according to Yu *et al.*[52] Thus, FINFETs will be good for low power CMOS applications. Joshi *et al.* performed simulation to show that the standby power of SRAM fabricated by FINFET technology is lower as shown in Fig. 4.29.[53] A review of FINFET transistor technology was given by Subramanian in 2010.[54]

Besides subthreshold swing S, GIDL is also important. Trivedi *et al.*[55] performed simulation on a FINFET structure and showed that "underlap" can reduce parasitic capacitance and also GIDL. Cho *et al.* performed a much more comprehensive simulation study on GIDL in FINFETs and pointed out that GIDL in FINFETs can

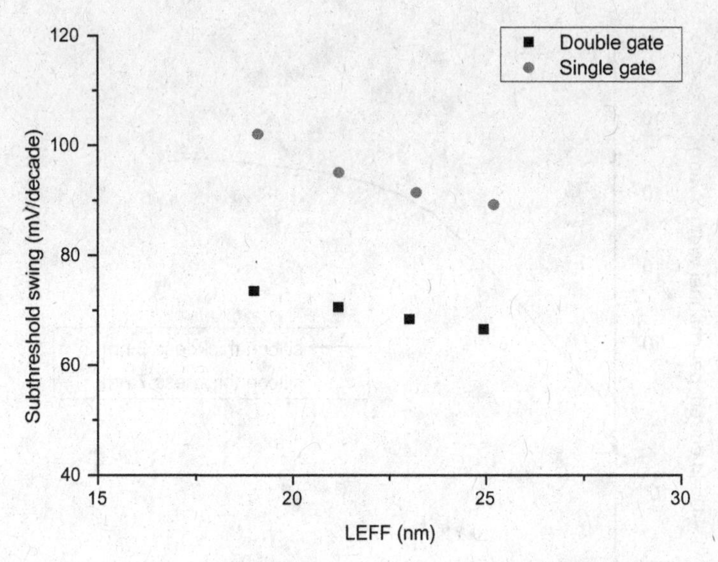

Fig. 4.28 Subthreshold swing S plotted against effective channel length for double gate (DG) and single gate bulk-silicon n-channel MOS transistors. (Modified from Fig. 2 in E. J. Nowak, I. Aller, T. Ludwig, K. Kim, R. V. Joshi, C.-T. Chuang, K. Bernstein and R. Puri, "Turning silicon on its edge: Overcoming silicon scaling barriers with double-gate and FinFET technology", *IEEE Circuits and Devices Magazine*, vol. 20, no. 1 (Jan./Feb. 2004), pp. 20–31.)

Table 4.3. FINFET performance (modified and simplified from Yu *et al.*[52])

	FINFET	FINFET
Lg (nm)	10	55
Vdd (V)	1.2	1.2
Tox (phys) (nm)	1.7	1.7
Gate material	Poly-Si	Poly-Si
Subthreshold swing for n-channel MOS transistor (mV/decade)	125	64
Subthreshold swing for p-channel MOS transistor (mV/decade)	101	68

be much smaller for FINFETs with larger "underlap length".[56] In conventional CMOS technology, there is always a positive "overlap" between the gate and drain (or source). In FINFET technology, it is possible to have a negative "overlap" between the gate and drain

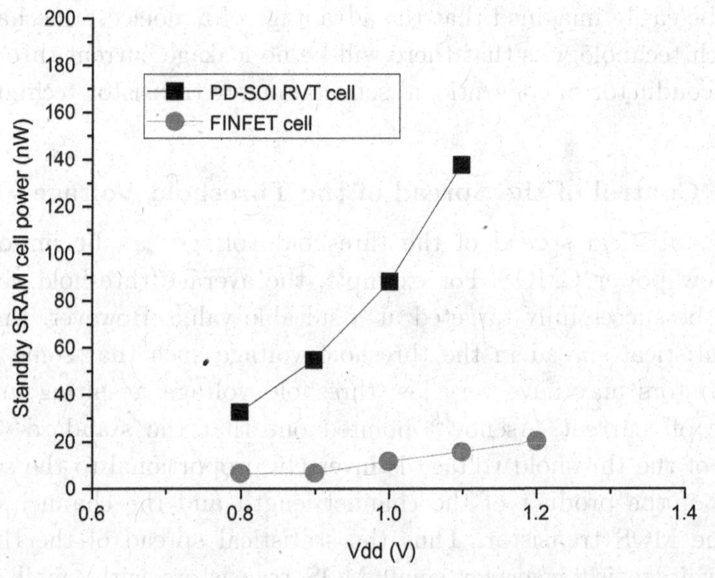

Fig. 4.29 Low standby power for SRAM using FINFET technology according to the simulation by Joshi *et al.*[53] (Modified from Fig. 11 in R. V. Joshi, R. Q. Williams, E. Nowak, K. Kim, J. Beintner, T. Ludwig, I. Aller and C. Chuang, "FinFET SRAM for high-performance low-power applications", *Proc. ESSDERC*, (2004), pp. 69–72.)

(or source). This negative "overlap" between the gate and drain (or source) has been known as "underlap". According to Yang *et al.*, FINFET technology with underlap can be "highly manufacturable" according to their simulation.[57]

4.4.8 *Reduction of power dissipation by mechanical switch technology*

In 2010, Theis and Solomon published a review paper regarding how to reduce power dissipation.[58] For example, the tunnel transistor was discussed. A review paper focused on the tunnel transistor has been given by Ionescu and Riel.[59] In 2012, Theis published a review paper regarding how to reduce power dissipation[60]; he mentioned the application of nanoelectromechanical switch technology to reduce power dissipation. This approach was also discussed by two articles from the research group of Professor T.-J. K. Liu (UC Berkerley).[61,62] It can be easily imagined that the advantage of nanoelectromechanical switch technology is that there will be no leakage current through a semiconductor in conventional semiconductor transistor technology.

4.5 Control of the Spread of the Threshold Voltage

The statistical spread of the threshold voltage can be important for low power CMOS. For example, the average threshold voltage may be successfully targeted at a suitable value. However, there is a statistical spread in the threshold voltage such that some MOS transistors may have very low threshold voltage, resulting in very large off current. Asenov[63] pointed out that the standard deviation of the threshold voltage is inversely proportional to the square root of the product of the channel length and the channel width of the MOS transistor. Thus the statistical spread of the threshold voltage is bigger for small MOS transistors with small channel length and/or channel width. In addition, Asenov[63] pointed out that the standard deviation of the threshold voltage is proportional to the square root of the dopant concentration in the channel region of the MOS transistor because of random dopant fluctuation (RDF). As shown in Fig. 4.30, the standard deviation of the threshold voltage appears to be proportional to $N^{0.4}$ according to Asenov and Saini[64]; N is the channel dopant concentration. Thus the statement that the standard deviation of the threshold voltage is proportional to the square root of the dopant concentration in the channel region of the MOS transistor is at least approximately correct according to experimental data.

Li *et al.*[65] pointed out that the statistical spread of the threshold voltage of planar MOS transistors using high-k metal gate technology is due to (1) process variation, (2) statistical spread of the work function of the metal gate and (3) random dopant fluctuation (RDF). Li *et al.*[66] pointed out that one way to reduce the statistical spread of the threshold voltage is to use two different gate metal materials for n-channel MOS transistors and p-channel MOS transistors and to use less dopants in n-channel MOS transistors and p-channel MOS transistors to reduce RDF.

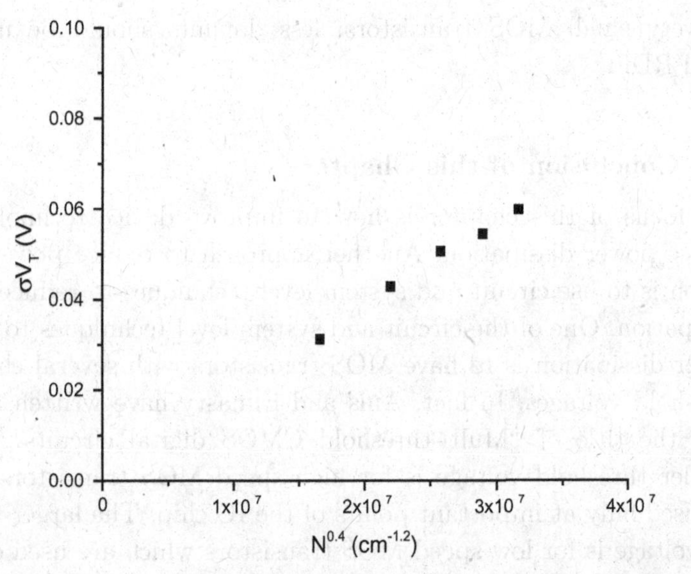

Fig. 4.30 The standard deviation of the threshold voltage plotted against the doping concentration to the power of 0.4 according to Asenov and Saini.[64] (Modified from Fig. 1 in A. Asenov and S. Saini, "Random dopant-induced threshold voltage fluctuations in sub-0.1-μm MOSFET's with epitaxial and δ-doped channels", *IEEE Trans. Electron Dev.*, vol. 46, no. 8 (Aug. 1999), pp. 1718–1724.)

The gate metal material for n-channel MOS transistors is known as n-metal while the gate metal material for p-channel MOS transistors is known as p-metal. One choice of n-metal is TiAlN [4.Cha] while one choice of the p-metal is TiN. There is a cap layer between the main high-k gate dielectric and the metal gate. The cap layer is also some sort of high-k dielectric. One choice of the cap layer for n-channel MOS transistor is La_2O_3 while one choice of the cap layer for p-channel MOS transistor is Al_2O_3. This can be seen, for example, in a review paper by Huang *et al.* [4.Huang]. The cap layer can also be used for threshold voltage control. The choice of the gate metal material and the cap layer can be used to control the threshold voltage; less dopants should be used. In Chapter 2, the author pointed out that the use of a threshold adjust implant was an important milestone in the history of MOS technology. However, the readers should note that the situation is somewhat different for very small MOS transistors; less dopants should be used to avoid RDF.

4.6 Conclusion of this Chapter

The focus of this chapter is how to improve device technology to reduce power dissipation. Another approach to reduce power dissipation is to use circuit and system level techniques to reduce power dissipation. One of the circuit and system level techniques to reduce power dissipation is to have MOS transistors with several choice of threshold voltages. In fact, Anis and Elmasry have written a book with the title of "Multi-threshold CMOS digital circuits".[69] The smaller threshold voltage is for high speed MOS transistors which are used only at important points of the IC chip. The larger threshold voltage is for low speed MOS transistors which are used only at non-critical points of the IC chip. Different threshold voltages can be achieved by using different channel lengths or different gate oxide thicknesses. This approach was reported, for example, by Sirisantana and Roy in 2004.[70] Another approach is to use multiple supply voltages. In 2004, Srivastava and Sylvester proposed to minimize total power by simultaneous V-dd/V-th assignment.[71] Subthreshold logic is to design a logic circuit to operate in the subthreshold regime, resulting in low speed but a big drop in power consumption.[72–76] Besides these methods, there are other approaches which are out of the scope of this book.

The technology to reduce power dissipation for specialized circuits can also be important and interesting. For example, low power SRAM technology has attracted a lot of attention. A lot of work has been done in the area of design of low power SRAM or even ultra-low power SRAM. For example, Mann *et al.* published a paper "Ultralow-power SRAM technology" in the IBM Journal of Research & Development in 2003.[77] As discussed above in this chapter, MOS transistors fabricated by UTB SOI or FINFET technology have very good subthreshold swing and so are very good for low power SRAM technology. A lot of work has been done in the area of design of low power microprocessors. For example, Piguet edited and published a book "Low-power processors and system on chips" in 2006.[78] Similarly, a lot of work has been done in the area of design of low power CMOS integrated circuits working at high frequencies. For example, Zolfaghari published a book "Low-power CMOS design for wireless transceivers" in 2003.[79]

Finally, the author would like to mention that there may exist other circuit and system level techniques to reduce power dissipation not mentioned in this book. Low power CMOS is still a very hot research area.

References

[1] R. Landauer, "Irreversibility and heat generation in the computing process", *IBM J. Res. Develop.*, vol. 5, no. 3 (July 1961), pp. 183–191. This article was re-printed in *IBM J. Res. Develop.*, vol. 44, no. 1/2 (January/March 2000), pp. 261–269. It was re-printed in the book *Maxwell's Demon: Entropy, Information, Computing*, edited by H. S. Leff and A. F. Rex, Princeton University Press, Princeton, New Jersey, 1990, pp. 188–196. In addition, it was also re-printed in the newer version of the above book "Maxwell's demon 2: entropy, classical and quantum information, computing", edited by H. S. Leff and A. F. Rex, IOP, Bristol and Philadelphia, 2003, pp. 148–156. Landauer's theory was also discussed in the book by van der Meer *et al.*: P. R. van der Meer, A. van Staveren and A. H. M. van Roermund, *Low-power Deep Sub-micron CMOS Logic: Sub-threshold Current Reduction*, Kluwer Academic Publishers, Boston, Massachusetts, USA, 2004, pp. 11–52.

[2] The power dissipation according to Landauer's mechanism is too small compared to the situation encountered by practical MOS integrated circuits. This was discussed in the book by Darling: D. Darling, *Teleportation: The Impossible Leap*, John Wiley & Sons, Hoboken, New Jersey, USA, 2005, pp. 121–122. Similarly, this was also discussed in the book by van der Meer et al.: P. R. van der Meer, A. van Staveren and A. H. M. van Roermund, *Low-Power Deep Sub-micron CMOS Logic: Sub-threshold Current Reduction*, Kluwer Academic Publishers, Boston, Massachusetts, USA, 2004, pp. 11–52.

[3] L. Chang, D. J. Frank, R. K. Montoye, S. J. Koester, B. L. Ji, P. W. Coteus, R. H. Dennard and W. Haensch, "Practical strategies for power-efficient computing technologies", *Proc. IEEE*, vol. 98, no. 2 (Feb. 2010), pp. 215–236.

[4] J. D. Meindl and R. N. Swanson, "Potential improvements in power-speed performanace of digital circuits", *Proc. IEEE*, vol. 59, no. 5 (May 1971), pp. 815–816.

[5] H. J. M. Veendrick, "Short-circuit power dissipation estimation for CMOS logic gates", *IEEE J. Solid-State Circuits*, vol. SC-19, no. 4 (Aug. 1984), pp. 468–473.

[6] S.-M. Kang and Y. Leblebici, *CMOS Digital Integrated Circuits: Analysis and Design*, Second Edition, McGraw-Hill, New York, 1999, p. 459.

[7] D. Auvergne, P. Maurine and N. Azemard, "Modeling for designing in deep submicron technologies", in *Low-power CMOS Circuits: Technology, Logic Design and CAD Tools*, edited by C. Piguet, Taylor & Francis, New York, 2006, pp. 6–8 to 6–10.

[8] P. R. van der Meer, A. van Staveren and A. H. M. van Roermund, *Low-power Deep Sub-micron CMOS Logic: Sub-threshold Current Reduction*, Kluwer Academic Publishers, Boston, Massachusetts, USA, 2004, pp. 1–154.

[9] Y. Tsukamoto, K. Nii, Y. Yamagami, T. Yoshizawa, S. Imaoka, T. Suzuki, A. Shibayama and H. Makino, "Comparison of the interconnect capacitances of various SRAM cell layouts to achieve high speed, low power SRAM cells", *Extended Abstracts of the 2003 International Conference on Solid State Devices and Materials (SSDM 2003)*, Tokyo, Japan, pp. 22–23.

[10] K. Imai, K. Yamaguchi, N. Kimizuka, H. Onishi, T. Kudo, A. Ono, K. Noda, Y. Goto, H. Fujii, M. Ikeda, K. Kazama, S. Maruyama, T. Kuwata and T. Horiuchi "A 0.13-μm CMOS technology integrating high-speed and low-power/high-density devices with two different well/channel structures", *IEDM Technical Digest*, (1999), pp. 667–670.

[11] H.-D. Lee and Y.-J. Lee, "Arsenic and phosphorus double ion implanted source/drain junction for 0.25- and sub-0.25-μm MOSFET technology", *IEEE Electron Dev. Lett.*, vol. 20, no. 1 (Jan. 1999), pp. 42–44.

[12] F.-C. Wang and C. Bulucea, "BF_2 and boron double implanted source/drain junctions for sub-0.25-μm CMOS technology", *IEEE Electron Dev. Lett.*, vol. 21, no. 10 (Oct. 2000), pp. 476–478.

[13] H. K. Kim, S. Y. Ong, E. Quek and S. Chu, "High performance device design through parasitic junction capacitance reduction and junction leakage current suppression beyond 0.1 μm technology", *Jpn. J. Appl. Phys.*, vol. 42, no. 4B (April 2003), pp. 2144–2148.

[14] J. W. Sleight, P. R. Varekamp, N. Lustig, J. Adkisson, A. Allen, O. Bula, X. Chen, T. Chou, W. Chu, J. Fitzsimmons, A. Gabor, S. Gates, P. Jamison, M. Khare, L. Lai, J. Lee, S. Narasimha, J. Ellis-Monaghan, K. Peterson, S. Rauch, S. Shukla, P. Smeys, T.-C. Su, J. Quinlan, A. Vayshenker, B. Ward, S. Womack, E. Barth, G. Biery, C. Davis, R. Ferguson, R. Goldblatt, E. Leobandung, J. Welser, I. Yang, and P. Agnello, "A high performance 0.13 μm SOI CMOS technology with a 70 nm silicon film and with a second generation low-k Cu BEOL," *IEDM Technical Digest*, (2001), pp. 245–248.

[15] G. G. Shahidi, "SOI technology for the GHz era", *IBM J. Res. & Dev.*, vol. 46, no. 2/3 (March/May 2002), pp. 121–131.

[16] H.-O. Joachim, Y. Yamaguchi, K. Ishikawa, Y. Inoue and T. Nishimura, "Simulation and two-dimensional analytical modeling of subthreshold slope in ultrathin-film SOI MOSFET's down to 0.1 μm gate length", *IEEE Trans. Electron Dev.*, vol. 40, no. 10 (Oct. 1993), pp. 1812–1817.

[17] Hans-Oliver Joachim, *Investigation on the Short-channel Silicon-on-insulator (SOI) MOSFET Towards 0.1 um Gate Length for Future VLSI Applications*", Shaker Verlag, Aachen, Germany, 1996.

[18] A. Pouydebasque, C. Charbuillet, R. Gwoziecki and T. Skotnicki, "Refinement of the subthreshold slope modeling for advanced bulk CMOS devices," *IEEE Trans. Electron. Dev.*, vol. 54, no. 10, pp. 2723–2729 (Oct. 2007).

[19] L. Pham-Nguyen, C. Fenouillet-Beranger, A. Vandooren, T. Skotnicki, G. Ghibaudo and S. Cristoloveanu, "In situ comparison of Si/high-k and Si/SiO_2 channel properties in SOI MOSFETs", *IEEE Electron Dev. Lett.*, vol. 30, no. 10 (Oct. 2009), pp. 1075–1077.

[20] K. Roy, S. Mukhopadhyay and H. Mahmoodi-Meimand, "Leakage current mechanisms and leakage reduction techniques in deep-submicrometer CMOS circuits", *Proc. IEEE*, vol. 91, no. 2 (Feb. 2003), pp. 305–327.

[21] A. Patel, "Minimize leakage power and process variations with dynamic V_t control", *Semiconductor International*, vol. 29, no. 11 (Oct. 2006), pp. 58–64.

[22] N. S. Kim, T. Austin, D. Blaauw, T. Mudge, K. Flautner, J. S. Hu, M. J. Irwin, M. Kandemire and V. Narayanan, "Leakage current: Moore's law meets static power", *Computer*, vol. 36, no. 12 (Dec. 2003), pp. 68–75.

[23] C. C. Wu, C. H. Diaz, B. L. Lin, S. Z. Chang, C. C. Wang, J. J. Liaw, C. H. Wang, K. K. Young, K. H. Lee, B. K. Liew and J. Y. C. Sun, "Ultra-low leakage 0.16 μm CMOS for low-standby power applications", *IEDM Technical Digest*, (1999), pp. 671–673.

[24] Y. Taur and T. H. Ning: *Fundamentals of Modern VLSI Devices* (Cambridge University Press, New York, 1998), p. 125.

[25] R. M. Swanson and J. D. Meindl, "Ion-implanted complementary MOS transistors in low-voltage circuits", *IEEE J. Solid-State Circuits*, vol. SC-7, no. 2 (April 1972), pp. 146–153.

[26] A. H.-C. Fung, "A subthreshold conduction model for BSIM", Electronics Research Laboratory, College of Engineering, University of California, Berkeley, Memorandum No. UCB/ERL M85/22, 1985.

[27] B. J. Sheu, D. L. Scharfetter, P.-K. Ko and M.-C. Jeng, "BSIM: Berkeley short-channel IGFET model for MOS transistors", *IEEE J. Solid-State Circuits*, vol. SC-22, no. 4 (August 1987), pp. 558–566.

[28] (A) J. R. Pfiester, J. D. Shott and J. D. Meindl, "Performance limits of CMOS ULSI", IEEE Trans. Electron Dev., vol. ED-32, no. 2 (Feb. 1985), pp. 333–343. (B) J. R. Pfiester, J. D. Shott and J. D. Meindl, "Performance limits of CMOS ULSI", *IEEE J. Solid-State Circuits*, vol. SC-20, no. 1 (Feb. 1985), pp. 253–263.

[29] J. R. Brews, "The submicron MOSFET", in *High-Speed Semiconductor Devices*, edited by S. M. Sze, John Wiley & Sons, 1990, pp. 139–209.

[30] K. Maitra and N. Bhat, "Poly reoxidation process step for suppressing edge direct tunneling through ultrathin gate oxides in NMOSFETs", *Solid-State Electron.*, vol. 47, no. 1 (Jan. 2003), pp. 15–17.

[31] J. Chen, T. Y. Chan, I. C. Chen, P. K. Ko and C. Hu, "Subbreakdown drain leakage current in MOSFET", *IEEE Electron Dev. Lett.*, vol. EDL-8, no. 11 (Nov. 1987), pp. 515–517.

[32] X. Yuan, J.-E. Park, J. Wang, E. Zhao, D. C. Ahlgren, T. Hook, J. Yuan, V. W. C. Chan, H. Shang, C.-H. Liang, R. Lindsay, S. Park and H. Choo, "Gate-induced-drain-leakage current in 45-nm CMOS technology," *IEEE Transactions on Device and Materials Reliability*, vol. 8, no. 3 (Sept. 2008), pp. 501–508.

[33] F. Gilibert, D. Rideau, A. Dray, F. Agut, M. Minondo, A. Juge, P. Masson and R. Bouchakour, "Characterization and modeling of gate-induced-drain-leakage", *IEICE Trans. Electron.*, vol. E88-C, no. 5 (May 2005), pp. 829–837.

[34] V. P.-H. Hu, M.-L. Fan, P. Su and C.-T. Chuang, "Band-to-band-tunneling leakage suppression for ultra-thin-body GeOI MOSFETs using transistor stacking", *IEEE Electron Dev. Lett.*, vol. 33, no. 2 (Feb. 2012), pp. 197–199.

[35] C. Y. Cheng, Y. K. Fang, J. C. Liao, T. J. Wang, Y. T. Hou, P. F. Hsu, K. C. Lin, K. T. Huang, T. L. Lee and M. S. Liang, "The effects of STI induced mechanical strain on GIDL current in Hf-based and SiON MOSFETs", *Solid-State Electron.*, vol. 53, no. 8 (August 2009), pp. 892–896.

[36] W. S. Lau, P. Yang, S. Y. Siah and L. Chan, "The role of a tensile stress bias for a sensitive silicon mechanical stress sensor based on a change in gate-induced-drain leakage current", *Microelectronics Reliability*, vol. 52, no. 11 (November 2012), pp. 2847–2850.

[37] S. E. Thompson, R. S. Chau, T. Ghani, K. Mistry, "In search of "Forever," continued transistor scaling one new material at a time," *IEEE Trans. Semi. Manuf.*, vol. 18, no. 1 (Feb. 2005), pp. 26–36. (Note: Similar theory can be found in a book chapter by S. Suthram, S. Narendra and S. Thompson, "Transistor design to reduce leakage", Chapter 12 in the book *Leakage in Nanometer CMOS Technologies*, edited by S. G. Narendra and A. Chandrakasan, Springer Science, New York, 2006, pp. 281–299.)

[38] D. Buss, "Device issues in the integration of analog/RF functions in deep submicron digital CMOS", *IEDM Technical Digest*, (1999), pp. 423–426.

[39] P. A. Stolk, H. P. Tuinhout, R. Duffy, E. Augendre, L. P. Bellefroid, M. J. B. Bolt, J. Croon, C. J. J. Dachs, F. R. J. Huisman, A. J. Moonen, Y. V. Ponomarev, R. F. M. Roes, M. Da Rold, E. Seevinck, K. N. Sreerambhatla, R. Surdeanu, R. M. D. A. Velghe, M. Vertregt, M. N. Webster, N. K. J. van Winkelhoff, and A. T. A. Zegers-Van Duijnhoven, "CMOS device optimization for mixed-signal technologies", *IEDM Technical Digest*, (2001), pp. 215–218.

[40] J.-C. Guo, "Halo and LDD engineering for multiple V_{TH} high performance analog CMOS devices", *IEEE Trans. Semi. Manuf.*, vol. 20, no. 3 (Aug. 2007), pp. 313–322.

[41] M.-C. Chang, C.-S. Chang, C.-P. Chao, K.-I. Goto, M. Ieong, L.-C. Lu and C. H. Diaz, "Transistor- and circuit-design optimization for low-power CMOS", *IEEE Trans. Electron Dev.*, vol. 55, no. 1 (Jan. 2008), pp. 84–95.

[42] W. S. Lau, P. Yang, E. T. L. Ng, Z. W. Chian, V. Ho, S. Y. Siah and L. Chan, "Region of nearly constant off current versus gate length characteristics for sub-0.1 μm low power CMOS technology", *Proc. IEEE EDSSC 2008*, pp. 231–234.

[43] W. S. Lau, P. Yang, E. H. Lim, Y. L. Tang, S. W. Lai, V. L. Lo, S. Y. Siah and L. Chan, "Observation of halo implant from the drain side reaching the source side and vice versa in extremely short p-channel transistors", *Microelectronics Reliability*, vol. 50, no. 3 (March 2010), pp. 346–350.

[44] P. Geens and W. Dehaene, "A small granular controlled leakage reduction system for SRAMs", *Solid-State Electron.*, vol. 49, no. 11 (Nov. 2005), pp. 1776–1782.

[45] C. Auth, "45nm high-k metal gate strained-enhanced CMOS transistors", *Proc. IEEE 2008 Custom Integrated Circuits Conference (CICC)*, pp. 379–386.

[46] J. Robertson and R. M. Wallace, "High-k materials and metal gates for CMOS applications", *Materials Science and Engineering R*, vol. 88 (2015), pp. 1–41.

[47] M. Gurfinkel, J. S. Suehle, J. B. Bernstein and Y. Shapira, "Enhanced gate induced drain leakage current in HfO_2 MOSFETs due to remote interface trap-assisted tunneling", *IEDM Technical Digest*, (2006), pp. 483–486. (Note: The page numbers may appear to be pp. 1–4 in IEEE database.)

[48] M. Gurfinkel, J. S. Suehle and Y. Shapira, "Enhanced gate induced drain leakage current in HfO_2 MOSFETs", *Microelectronic Engineering*, vol. 86, no. 11 (Nov. 2009), pp. 2157–2160.

[49] J. C. Liao, Y.-K. Fang, Y. T. Hou, W. H. Tseng, P. F. Hsu, K. C. Lin, K. T. Huang, T. L. Lee and M. S. Liang, "Investigation of bulk traps enhanced gate-induced leakage current in Hf-based MOSFETs", *IEEE Electron Dev. Lett.*, vol. 29, no. 5 (May 2008), pp. 509–511.

[50] Y.-K. Choi, K. Asano, N. Lindert, V. Subramanian, T.-J. King, J. Bokor and C. Hu, "Ultrathin-body SOI MOSFET for deep-sub-tenth micron era", *IEEE Electron Dev. Lett.*, vol. 21, no. 5 (May 2000), pp. 254–255.

[51] E. J. Nowak, I. Aller, T. Ludwig, K. Kim, R. V. Joshi, C.-T. Chuang, K. Bernstein and R. Puri, "Turning silicon on its edge: Overcoming silicon scaling barriers with double-gate and FinFET technology", *IEEE Circuits and Devices Magazine*, vol. 20, no. 1 (Jan./Feb. 2004), pp. 20–31.

[52] B. Yu, L. Chang, S. Ahmed, H. Wang, S. Bell, C.-Y. Yang, C. Tabery, C. Ho, Q. Xiang, T.-J. King, J. Bokor, C. Hu, M.-R. Lin and D. Kyser, "FINFET scaling to 10nm gate length", *IEDM Technical Digest*, (2002), pp. 251–254.

[53] R. V. Joshi, R. Q. Williams, E. Nowak, K. Kim, J. Beintner, T. Ludwig, I. Aller and C. Chuang, "FinFET SRAM for high-performance low-power applications", *Proc. ESSDERC*, (2004), pp. 69–72.

[54] V. Subramanian, "Multiple gate field-effect transistors for future CMOS technologies", *IETE Technical Review*, vol. 27, no. 6 (Nov.–Dec. 2010), pp. 446–454.

[55] V. Trivedi, J. G. Fossum and M. M. Chowdhury, "Nanoscale FinFETs with gate-source/drain underlap," *IEEE Trans. Electron. Dev.*, vol. 52, no. 1 (June 2005), pp. 56–62.

[56] S. Cho, S. O'uchi, K. Endo, T. Matsukawa, K. Sakamoto, Y. Liu, B.-G. Park and M. Masahara, "Minimization of gate-induced drain leakage by controlling gate underlap length for low-standby-power operation of 20-nm-level four-terminal silicon-on-insulator Fin-shaped field effect transistor", *Jpn. J. Appl. Phys.*, vol. 49, no. 2 issue 1 (February 2010), pp. 024203-1 to 024203-5.

[57] J.-W. Yang, P. M. Zeitzoff and H.-H. Tseng, "Highly manufacturable double-gate FinFET with gate-source/drain underlap," *IEEE Trans. Electron. Dev.*, vol. 54, no. 6 (June 2007), pp. 1464–1470.

[58] T. N. Theis and P. M. Solomon, "In quest of the next switch: Prospects for greatly reduced power dissipation in a successor to the silicon field-effect transistor", *Proc. IEEE*, vol. 98, no. 12 (Dec. 2010), pp. 2005–2014.

[59] A. M. Ionescu and H. Riel, "Tunnel field-effect transistors as energy-efficient electronic switches", *Nature*, vol. 479, no. 7373 (17 November 2011), pp. 329–337.

[60] T. N. Theis, "In quest of a fast, low-voltage digital switch", *ECS Transactions*, vol. 45, no. 6 (2012), pp. 3–11.

[61] V. Pott, H. Kam, R. Nathanael, J. Jeon, E. Alon and T.-J. K. Liu "Mechanical computing redux: relays for integrated circuit applications", *Proc. IEEE*, vol. 98, no. 12 (Dec. 2010), pp. 2076–2094.

[62] T.-J. K. Liu, D. Markovic, V. Stojanoic and E. Alon, "The relay reborn", *IEEE Spectrum*, vol. 49, no. 4 (April 2012), pp. 40–43.

[63] A. Asenov, "Random dopant induced threshold voltage lowering and fluctuations in sub-0.1 μm MOSFET's: A 3-D "Atomistic" simulation study", *IEEE Trans. Electron Dev.*, vol. 45, no. 12 (Dec. 1998), pp. 2505–2513.

[64] A. Asenov and S. Saini, "Random dopant-induced threshold voltage fluctuations in sub-0.1-μm MOSFET's with epitaxial and δ-doped channels", *IEEE Trans. Electron Dev.*, vol. 46, no. 8 (Aug. 1999), pp. 1718–1724.

[65] Y. Li, C.-H. Hwang, T.-Y. Li and M.-H. Han, "Process-variation effect, metal-gate work-function fluctuation, and random-dopant fluctuation in emerging CMOS technologies", *IEEE Trans. Electron Dev.*, vol. 57, no. 2 (Feb. 2010), pp. 437–447.

[66] Y. Li, K.-F. Lee, C.-Y. Yiu, Y.-Y. Chiu and R.-W. Chang, "Dual-material gate approach to suppression of random-dopant-induced characteristic fluctuation in 16 nm metal-oxide-semiconductor field-effect-transistor devices", *Jpn. J. Appl. Phys.*, vol. 50 (2011), article number 04DC07.

[67] T.-H. Cha, D.-G. Park, T.-K. Kim, S.-A. Jang, I.-S. Yeo, J.-S. Roh and J. W. Park, "Work function and thermal stability of $Ti_{1-x}Al_xN_y$ for dual metal gate electrodes", *Appl. Phys. Lett.*, vol. 81, no. 22 (25 Nov. 2002), pp. 4192–4194.

[68] Huang AnPing, Zheng XiaoHu, Xiao ZhiSong, Wang Mei, Di ZengFeng and Chu Paul K., “Interface dipole engineering in metal gate/high-k stacks”, *Chinese Science Bulletin*, vol. 57, no. 22 (August 2012), pp. 2872–2878. (Note: This paper originates from China and the authors put their surname in front of the given name in the the authors' list.)

[69] M. Anis and M. Elmasry, *Multi-threshold CMOS Digital Circuits: Managing Leakage Power*, Kluwer Academic Publishers, Norwell, Massachusetts, USA, 2003, pp. 1–216.

[70] N. Sirisantana and K. Roy, “Low-power design using multiple channel lengths and oxide thicknesses,” *IEEE Design & Test of Computers*, vol. 21, no. 1 (Jan.–Feb. 2004), pp. 56–63.

[71] A. Srivastava and D. Sylvester, “Minimizing total power by simultaneous V-dd/V-th assignment,” *IEEE Transactions on Computer-aided Design of Integrated Circuits and Systems*, vol. 23, no. 5 (May 2004), pp. 665–677.

[72] H. Soeleman, K. Roy and B. C. Paul, “Robust subthreshold logic for ultra-low power operation,” *IEEE Trans. VLSI Systems*, vol. 9, no. 1 (Feb. 2001), pp. 90–99.

[73] B. C. Paul, A. Raychowdhury and K. Roy, “Device optimization for digital subthreshold logic operation,” *IEEE Trans. Electron Dev.*, vol. 52, no. 2 (Feb. 2005), pp. 237–247.

[74] B. H. Calhoun, A. Wang and A. Chandrakasan, “Modeling and sizing for minimum energy operation in subthreshold circuits,” *IEEE Journal of Solid-State Circuits*, vol. 40, no. 9 (Sep. 2005), pp. 1778–1786.

[75] S. Hanson, S. Mingoo, D. Sylvester and D. Blaauw, “Nanometer device scaling in subthreshold logic and SRAM,” *IEEE Trans. Electron Dev.*, vol. 55, no. 1 (Jan. 2008), pp. 175–185.

[76] B. C. Paul and K. Roy, “Oxide thickness optimization for digital subthreshold operation,” *IEEE Trans. Electron Dev.*, vol. 55, no. 2 (Feb. 2008), pp. 685–688.

[77] R. W. Mann, W. W. Abadeer, M. J. Breitwisch, O. Bula, J. S. Brown, B. C. Colwill, P. E. Cottrell, W. G. Crocco Jr., S. S. Furkay, M. J. Hauser, T. B. Hook, D. Hoyniak, J. M. Johnson, C. H. Lam, R. D. Mih, J. Rivard, A. Moriwaki, E. Phipps, C. S. Putnam, B. A. Rainey, J. J. Toomey and M. I. Younus, “Ultralow-power SRAM technology”, *IBM J. Res. & Dev.*, vol. 47, no. 5/6 (September/November 2003), pp. 553–566.

[78] *Low-power Processors and System on Chips*, edited by C. Piguet, Taylor & Francis, New York, 2006, pp. 1–1 to 20–24.

[79] A. Zolfaghari, Low-power CMOS Design for Wireless Transceivers, Kluwer Academic Publishers, Boston, 2003, pp. 1–106.

Chapter Five

Analog CMOS Technology

In the beginning, CMOS technology was used for digital integrated circuits. Subsequently, CMOS technology was also used to make analog integrated circuits. For example, operational amplifiers can be made by CMOS.

5.1 Low Frequency Analog CMOS

5.1.1 *Significance of the Early voltage*

Besides digital integrated circuits, CMOS technology can also be used to make analog integrated circuits like operational amplifiers.[1,2] The analysis of the voltage gain of an amplifier based on MOS transistors used in an analog integrated circuit is different from the analysis of the voltage gain of an amplifier based on MOS transistors used in an electronic circuit based on discrete components. An important factor is that a resistor can be very wasteful of silicon area such that resistors are avoided in the design of analog integrated circuits.

The voltage gain of an amplifier based on an MOS transistor with a resistive load R_L is given by $-g_m R_L$, where g_m is the transconductance of the MOS transistor. For sinusoidal input, the negative sign shows that there is a 180° phase shift. If only the magnitude is concerned, the magnitude of the voltage gain of an amplifier based on an MOS transistor with a resistive load R_L is given by $g_m R_L$. However, in analog CMOS, an MOS transistor is quite frequently used as a dynamic load for another MOS transistor such that the magnitude of the gain is given by the following equation:

$$A_v = g_m/g_d \tag{5.1}$$

In Eq. (5.1), the drain conductance g_d is given by the derivative of the drain current with respect to the drain voltage. Thus $g_d = dI_d/dV_d$, where I_d and V_d are the drain current and drain voltage respectively. (Note: In strict sense, g_d in Eq. (5.1) should be the parallel combination of the g_d due to the MOS transistor serving to provide gain and the g_d due to the MOS transistor serving as the dynamic load.) In the saturation region where the drain current increases slowly with drain voltage, g_d is given approximately by the following equation.

$$g_d = I_d/V_A \tag{5.2}$$

Subsitution of Eq. (5.2) into Eq. (5.1) results in Eq. (5.3) as follows.

$$A_v = (g_m/I_d)V_A \tag{5.3}$$

V_A is known as the Early voltage. The Early voltage was named after James M. Early (1922–2004). When he was a scientist working in the Bell Telephone Laboratories, Early published a famous paper on bipolar transistors with the title "Effects of space-charge layer widening in junction transistors" in 1952.[3] The readers should note that for bipolar transistors, I_d should be replaced by the collector current I_c in Eq. (5.3) such that $A_v = (g_m/I_c)V_A$. The same concept was subsequently extended to MOS transistors, thus the concept of the Early voltage is applicable to both bipolar transistors and MOS transistors. As seen from Eq. (5.3), higher Early voltage will lead to higher voltage gain.

In reality, the amplifier circuit may be more complicated than a simple MOS transistor with another MOS transistor serving as a dynamic load. For example, the amplifier circuit may be a differential pair of MOS transistors with some sort of current mirror serving as the load, as shown in Fig. 5.1. However, in general, the concept that higher Early voltage will lead to higher voltage gain is still valid.

In Table 5.1 (modified from Table I of Kilchytska *et al.* 2003),[4] it can be easily seen that the use of halo implant in conventional bulk CMOS technology, the Early voltage is degraded for long channel

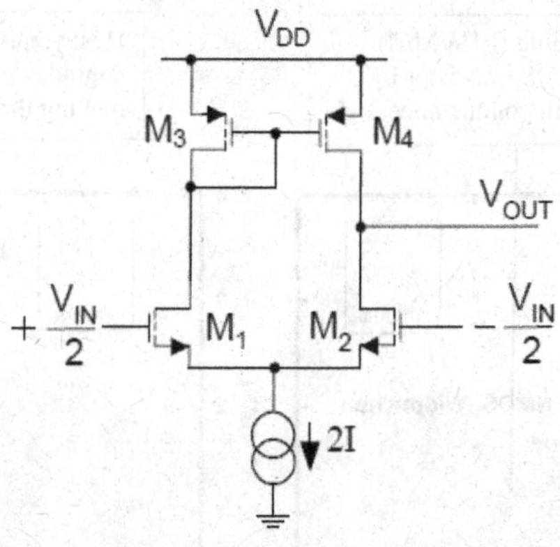

Fig. 5.1 A differential amplifier with two n-channel MOS transistors forming a differential pair and two p-channel MOS transistors forming a current mirror which serves as a dynamic load for the two n-channel transistors. Both the Early voltage of the n-channel transistor and the Early voltage of the p-channel transistor will affect the voltage gain of the differential amplifier. Higher Early voltage will lead to higher voltage gain.

Table 5.1. Early voltage (V) for MOS transistors with and without halo implant.

Channel length (um)	Bulk CMOS with halo	Bulk CMOS without halo	Fully depleted SOI CMOS with halo	Fully depleted SOI CMOS without halo
0.08	1.12	0.69	0.67	0.75
0.1	1.30	0.76	0.72	0.83
0.12	1.73	0.89	0.86	0.92
0.15	2.01	1.10	1.03	1.18
0.2	2.98	1.78	1.63	1.88
0.3	3.23	2.04	2.83	
0.5	4.35	9.70	2.82	4.69
1	5.17	18.85	4.94	11.06

devices. For example, for a channel length of 1 μm, the Early voltage is degraded from 18.85 V to 5.17 V when halo implant is used. Similarly, in Table 5.1, it can be easily seen that the use of halo implant in SOI CMOS technology, the Early voltage is also degraded for long channel devices. For example, for a channel length of 1 μm, the Early voltage is degraded from 11.06 V to 4.94 V when halo

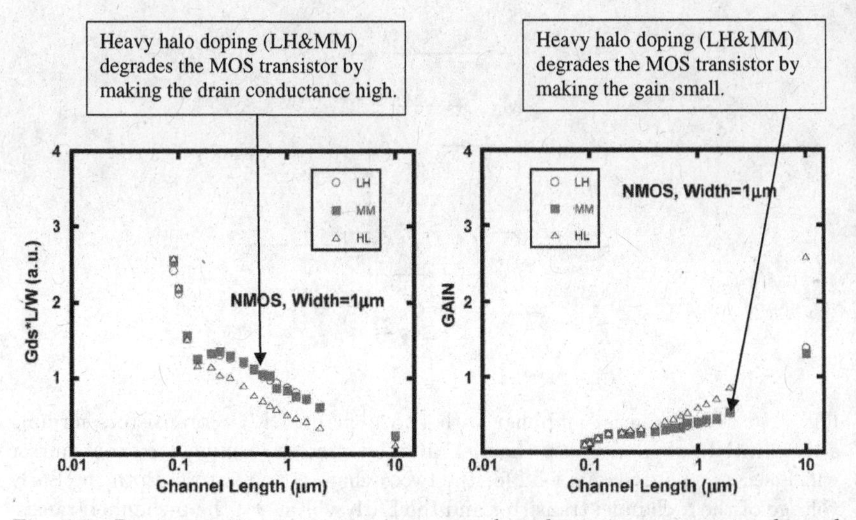

Fig. 5.2 Representative output resistance and analog-gain response to channel and pocket profile engineering balance in an advanced CMOS technology according to Chang *et al.*[5] L, M, H stand for low, medium and high respectively. LH, MM and HL stand for the split condition with light channel & heavy halo doping, the split condition with medium channel & medium halo doping and the split condition with heavy channel & light halo doping. (Modified from Fig. 19 in M.-C. Chang, C.-S. Chang, C.-P. Chao, K.-I. Goto, M. Ieong, L.-C. Lu and C. H. Diaz, "Transistor- and circuit-design optimization for low-power CMOS", *IEEE Trans. Electron Dev.*, vol. 55, no. 1 (Jan. 2008), pp. 84–95.)

implant is used for fully depleted (FD) SOI CMOS technology. In 2008, Chang *et al.* published a paper which included a discussion on the MOS transistor gain.[5] Figure 5.2 shows that the MOS amplifier gain becomes larger when the channel length becomes larger. For an MOS transistor amplifier with a resistor serving as the load, the MOS amplifier gain is expected to become smaller when the channel length becomes larger because of a decrease in the transconductance for large channel length. However, the situation can be different for an MOS transistor amplifier with another MOS transistor serving as a dynamic load. Figure 5.2 shows that the MOS amplifier gain becomes smaller when halo implant dose becomes larger. As discussed in earlier chapters, the halo implant helps to suppress short channel effect and also punchthrough; however, the halo implant can degrade the performance of analog CMOS as discussed above in this paragraph.

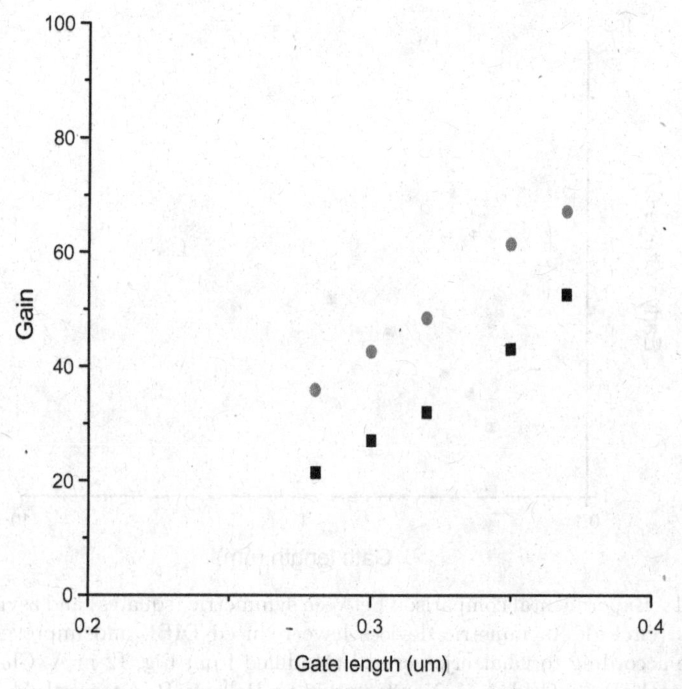

Fig. 5.3 The gain g_m/g_d of n-channel MOS transistors with (circles) and without halo implant (squares) according to Guo 2007.[6] The data show that the gain can be larger due to the halo implant for short MOS transistors. (Modified from Fig. 1 in J.-C. Guo, "Halo and LDD engineering for multiple Vth high performance analog CMOS devices", *IEEE Trans. Semiconductor Manufacturing*, vol. 20, no. 3 (Aug. 2007), pp. 313–322.)

The readers should note that the halo implant can degrade the gain of long MOS transistors. However, this may not be the case for short MOS transistors. As shown in Fig. 5.3, the gain can be larger due to the halo implant for short MOS transistors according to Guo.[6]

Traditionally, a symmetric halo implant design is used in short MOS transistors with a halo implant for the drain and also a halo implant for the source. In 1996, Miyamoto *et al.* proposed an asymmetric halo implant design with a halo implant for the source only.[7] Figure 5.4 shows that the Early voltage of MOS transistors with an asymmetric halo implant design is significantly larger than that of MOS transistors with a symmetric halo implant design according to

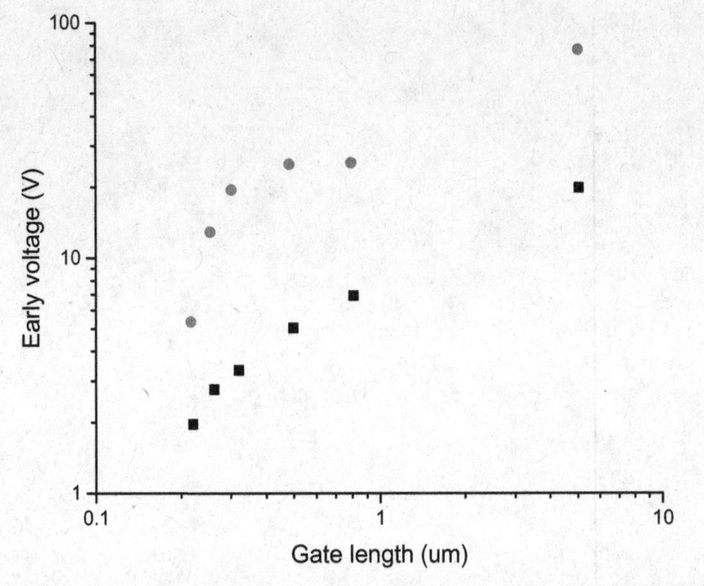

Fig. 5.4 Experimental comparison between symmetric (squares) and asymmetric devices (circles). Asymmetric devices have reduced DIBL and improved Early voltage according to Chatterjee *et al.*[8] (Modified from Fig. 12 in A. Chatterjee, K. Vasanth, D. T. Grider, M. Nandakumar, G. Pollack, R. Aggarwal, M. Rodder and H. Shichijo, "Transistor design issues in integrating analog functions with high performance digital CMOS", Symposium on VLSI Technology Digest of Technical Papers, pp. 147–148, 1999.)

Chatterjee *et al.*[8] In addition, Fig. 5.4 also shows that the DIBL (drain induced barrier lowering) of MOS transistors with an asymmetric halo implant design is significantly smaller than that of MOS transistors with a symmetric halo implant design. (Note: DIBL is quite frequently measured in terms of the difference between the linear threshold voltage and the saturation threshold voltage.) The interested readers should note that the DIBL and the Early voltage are related. In general, a large DIBL implies a small Early voltage whereas a small DIBL implies a large Early voltage. A small DIBL and a large Early voltage are usually desirable for an MOS transistor.

Figure 5.4 shows DIBL plotted against gate length; Fig. 5.5 shows DIBL plotted against the drain voltage for MOS transistors

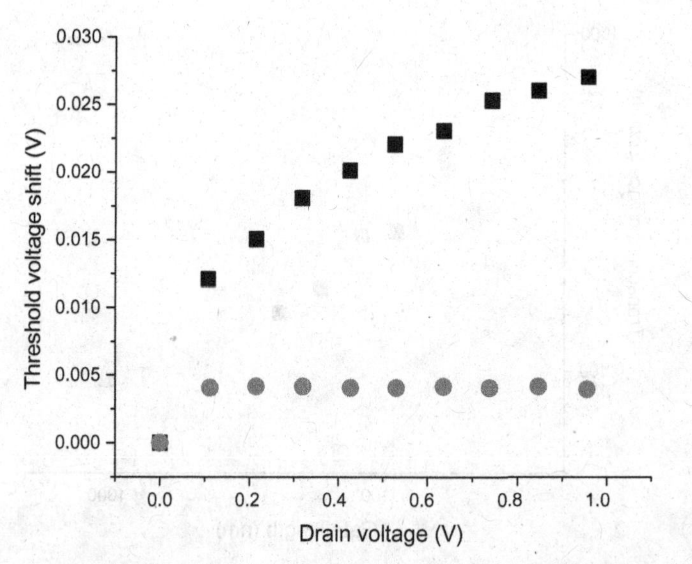

Fig. 5.5 Advantage of the choice of an asymmetric halo implant design in the reduction of the threshold voltage shift (DIBL) according to Roy *et al.*[9] (Modified from Fig. 1 in A. S. Roy, S. P. Mudanai and M. Stettler, "Mechanism of long-channel drain-induced barrier lowering in halo MOSFETs", *IEEE Trans. Electron Dev.*, vol. 58, no. 4 (Apr. 2011), pp. 979–984.)

with asymmetric halo and symmetric halo according to Roy *et al.*[9] Figure 5.5 also shows that the DIBL (drain induced barrier lowering) of MOS transistors with an asymmetric halo implant design is significantly smaller than that of MOS transistors with a symmetric halo implant design.

The mechanism of degradation of the performance of MOS transistors by the halo implant has been discussed by various authors. For example, in 1999, Cao *et al.*[10] and Buss discussed the mechanism of this issue.[11] Two years later in 2001, Stolk *et al.* also discussed the mechanism of this issue.[12] Thus the degradation of MOS transistors for analog CMOS integrated circuits by the halo implant has been considered an important issue by various authors.

As discussed in Chapter Four, FINFETs have superior gate control of channel compared to conventional planar bulk MOS transistors. In other words, FINFETs have inferior drain control of channel compared to conventional planar bulk MOS transistors. Thus it is

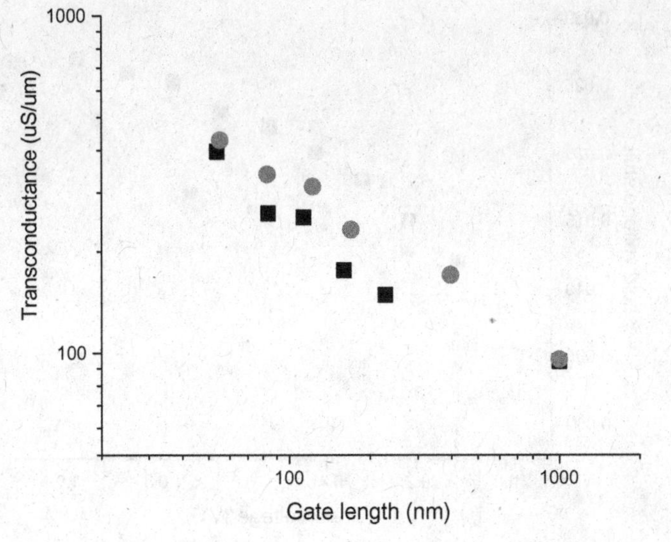

Fig. 5.6 The transconductance g_m at $V_{DS} = 1.2\,V$ and $V_{GS} - V_T = 0.2\,V$ according to Subramanian *et al.*[13] The transconductance g_m becomes bigger with the decrease of gate length. However, it is nearly the same for planar bulk MOS transistors or FINFETs.

expected that FINFETs have smaller drain induced barrier lowering (DIBL) compared to conventional planar bulk MOS transistors, resulting in higher Early voltage V_A and smaller output conductance g_{ds}. Subramanian *et al.*[13] made a detailed comparison of FINFETs and conventional planar bulk MOS transistors for analog/RF applications. As shown in Fig. 5.6, the transconductance g_m becomes bigger with the decrease of gate length; however, it is nearly the same for planar bulk MOS transistors or FINFETs. As shown in Fig. 5.7, FINFETs have lower (i.e. better) g_{ds} than planar bulk MOS transistors due to superior gate control of channel. As shown in Fig. 5.8, a similar g_m and reduced g_{ds} translate into higher voltage gains for FINFETs compared to planar bulk MOS transistors.

5.1.2 *Significance of transistor matching*

In 2009, Lewyn *et al.* briefly discussed the significance of transistor mismatch in analog CMOS technologies.[14] In analog CMOS technology, differential amplifiers or operational amplifiers are frequently

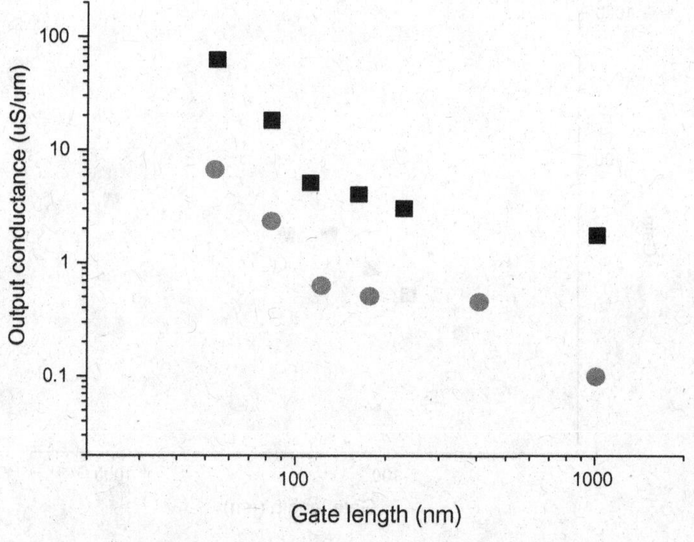

Fig. 5.7 The output conductance g_{ds} at $V_{DS} = 1.2\,V$ and $V_{GS} - V_T = 0.2\,V$ according to Subramanian *et al.*[13] FINFETs have lower (i.e. better) g_{ds} than planar bulk MOS transistors due to superior gate control of channel.

used; for this sort of amplifiers, MOS transistor matching is important. In the book written by Croon *et al.* in 2005, the authors pointed out that good transistor matching is required to make high accuracy analog-to-digital converters (ADCs) with acceptable yield.[15] In the book written by Marshall in 2009, the author pointed out besides analog circuits like the current mirror and operational amplifier, a digital circuit like SRAM can also be sensitive to MOS transistor mismatch.[16]

In 1986, Lakshmikumar *et al.* published a paper on MOS transistor matching.[17] In 1988, Conroy *et al.*[18] published an article to comment on the paper by Lakshmikumar *et al.* In 1988, Pelgrom *et al.* published a preliminary paper on MOS transistor matching[19]; subsequently in 1989, Pelgrom *et al.* published a full paper on MOS transistor matching.[20] It appears to the author that the work of Pelgrom *et al.* has really a strong impact on the issue of MOS transistor matching; up to 2014, the full paper published in 1989 has more than 2000 citations.

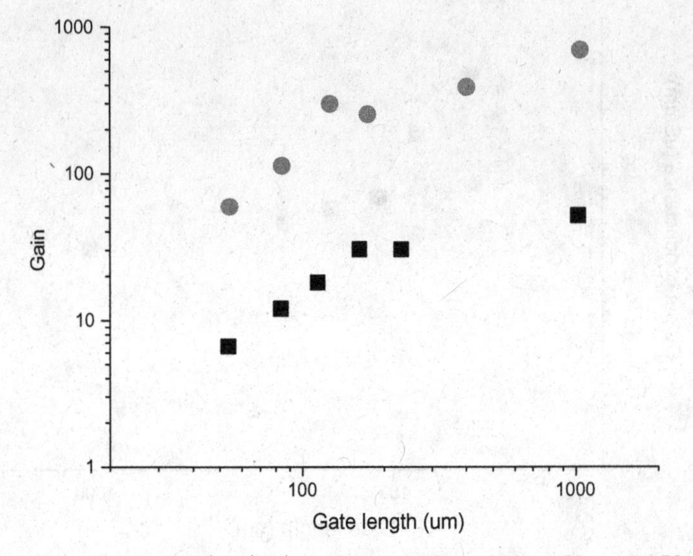

Fig. 5.8 The voltage gain (g_m/g_{ds}) at $V_{DS} = 1.2\,V$ and $V_{GS} - V_T = 0.2\,V$ according to Subramanian *et al.*[13] A similar g_m and reduced g_{ds} translate into higher voltage gains for FINFETs compared to planar bulk MOS transistors. (Modified from Fig. 5, Fig. 6 and Fig. 7 in V. Subramanian, B. Parvais, J. Borremans, A. Mercha, D. Linten, P. Wambacq, J. Loo, M. Dehan, C. Gustin, N. Collaert, S. Kubicek, R. Lander, J. Hooker, F. Cubaynes, S. Donnay, M. Jurczak, G. Groeseneken, W. Sansen and S. Decoutere, "Planar bulk MOSFETs versus FinFETs: An analog/RF perspective", *IEEE Trans. Electron Dev.*, vol. 53, no. 12 (Dec. 2006), pp. 3071–3079.)

MOS transistors with a homogeneously doped channel follow the Pelgrom MOS transistor matching model as follows. The square of the standard deviation of threshold voltage is given by

$$\sigma^2_{Vt} = A^2_{Vt}/(WL) \tag{5.4}$$

where W and L are the gate width and gate length respectively while A_{Vt} is the matching parameter.

In the older days when halo implant was not used to suppress short channel effect A_{Vt} was more or less a constant for different gate lengths and gate widths. In more advanced CMOS technology, halo implant is used to suppress short channel effect and A_{Vt} is no longer a constant for different gate lengths and gate widths. According to Chang *et al.*,[5] it is still possible to define IA_{Vt} for an individual device pair with a particular gate length L_1 and a particular gate width W_1 as follows:

$$\sigma_{Vt}^2 = IA_{Vt}^2/(W_1L_1) \quad (5.5)$$

$$IA_{Vt} = \sigma_{Vt}(W_1L_1)^{1/2} \quad (5.6)$$

IA_{Vt} is an individual device pair mismatch parameter. Figure 5.9 shows a conventional Pelgrom plot (a plot of the standard variation of the threshold voltage against the reciprocal of the square root of the product of gain width and gate length) for n-channel MOS

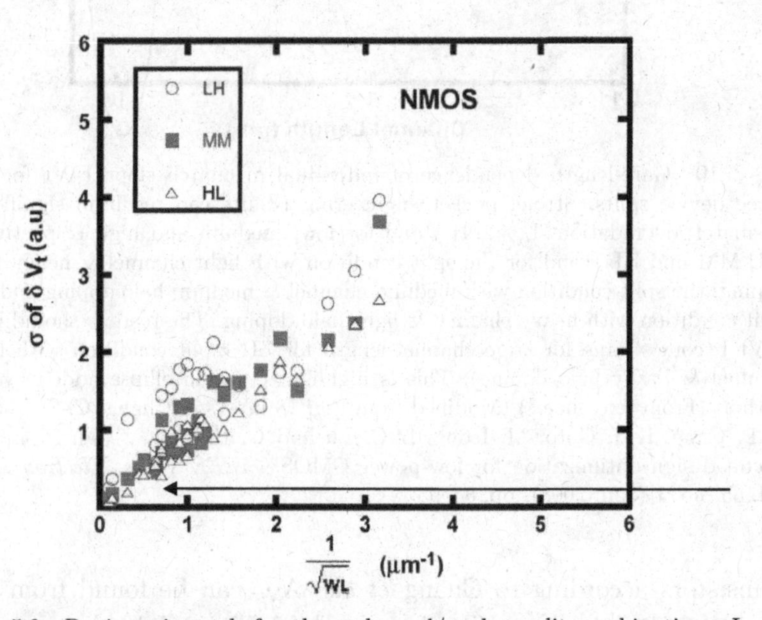

Fig. 5.9 Device mismatch for three channel/pocket split combinations. L, M, H stand for low, medium and high respectively. LH, MM and HL stand for the split condition with light channel & heavy halo doping, the split condition with medium channel & medium halo doping and the split condition with heavy channel & light halo doping. An arrow was added by the author to show that this figure does not show any significant difference for the 3 split conditions for large gate lengths. However, Fig. 5.10 shows that the 3 split conditions show big difference for large gate lengths. (From reference. 5) (Modified from Fig. 17 in M.-C. Chang, C.-S. Chang, C.-P. Chao, K.-I. Goto, M. Ieong, L.-C. Lu and C. H. Diaz, "Transistor- and circuit-design optimization for low-power CMOS", *IEEE Trans. Electron Dev.*, vol. 55, no. 1 (Jan. 2008), pp. 84–95.)

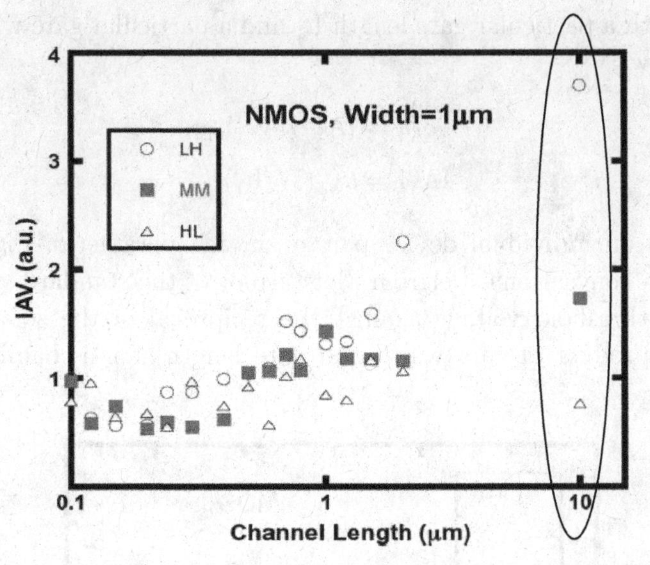

Fig. 5.10 Gate-length dependence of individual mismatch slope IAVt for the three device splits. Strong pocket engineering results can result in significant mismatch degradation. L, M, H stand for low, medium and high respectively. LH, MM and HL stand for the split condition with light channel & heavy halo doping, the split condition with medium channel & medium halo doping and the split condition with heavy channel & light halo doping. The readers should note IAVt becomes large for large channel length for LH (split condition with light channel & heavy halo doping). This is highlighted by an ellipse added by the author. (From reference.[5]) (Modified from Fig. 18 in M.-C. Chang, C.-S. Chang, C.-P. Chao, K.-I. Goto, M. Ieong, L.-C. Lu and C. H. Diaz, "Transistor- and circuit-design optimization for low-power CMOS", *IEEE Trans. Electron Dev.*, vol. 55, no. 1 (Jan. 2008), pp. 84–95.)

transistors according to Chang *et al.*[5] A_{Vt} can be found from the slope if a straight line is fitted to the data. Figure 5.10 shows a plot of IA_{Vt} against the gate length for a fixed gate width for n-channel MOS transistors according to Chang *et al.*[5] The mismatch of MOS transistors with halo implant was further discussed by Yuan *et al.*[21] in 2011. A model of the mismatch of MOS transistors with halo implant was proposed by Schaper and Einfeld[22] in 2011.

It is known that FINFETs perform better than conventional planar MOS transistors[23] in terms of transistor matching. There are two possible mechanisms. One mechanism is the excellent short channel performance of FINFETs; two MOS transistors manufactured by conventional planar MOS technology may have a small difference in threshold voltage because the two MOS transistors may have slightly different channel lengths. (Note: FINFETs have superior gate control of channel or inferior drain control of control, resulting in a small "short channel effect". Short channel effect has been discussed in Chapter Two.) Another mechanism is that FINFET technology is quite frequently used together with high-k metal gate technology such that FINFETs quite frequently do not depend on threshold adjust implant for the setting of the threshold voltage; the tuning of the threshold voltage can be achieved by tuning the work function of the gate metal. Lower channel doping can lead to less statistical variation in the threshold voltage because of less random dopant fluctuation (RDF).

5.2 High Frequency Analog CMOS

5.2.1 *Low noise RF amplifier*

The noise figure F is given by

$$F = 1 + (R_G/R_S) + \gamma g_{do} R_S (\omega/\omega_T)^2 \tag{5.7}$$

R_S is the internal resistance of the RF signal source. R_G is the gate resistance of the MOS transistor. In addition, g_{do} is the zero-bias conductance of the device. γ is a bias dependent factor of the device with the value between 2/3 and 1 and ω_T is the angular cut-off frequency approximately equal to $g_m C_{gs}$. Equation (5.7) comes from Eq. (13) in Shaeffer and Lee 1997.[24] (Note: Shaeffer and Lee published a short correction of their 1997 work later in 2005.)[25]

For lower noise, F should be made smaller! It can be easily seen that ω_T is made larger, F will be smaller. As sown in Fig. 5.11, f_T (i.e. $\omega_T/2\pi$) becomes larger when the gate length becomes smaller.[26] According to Eq. (5.7), the reduction of the noise figure F can be done by using smaller gate length. However, F may become larger because the gate resistance R_G becomes larger because of the smaller gate length. This problem can be solved by using multiple fingers for the gate. If the problem of larger gate resistance is solved, the

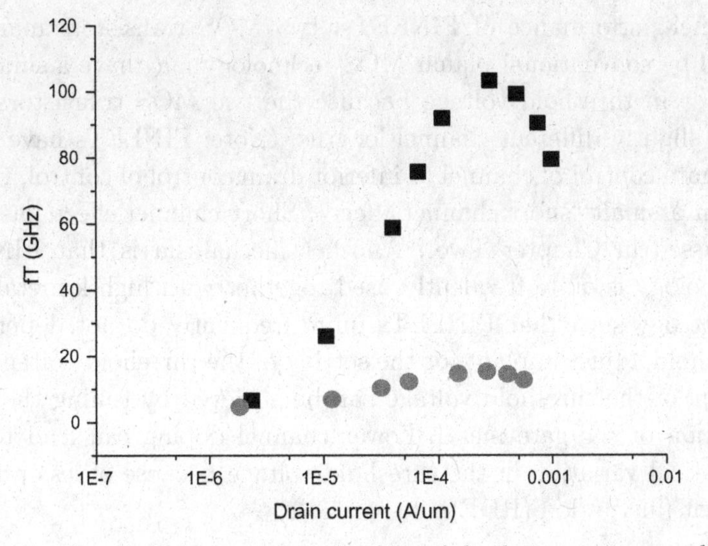

Fig. 5.11 The cut-off frequency as a function of the drain current according to Iwai and Momose.[26] The plot for a gate length of 0.07 μm is in squares whereas the plot for a gate length of 0.51 μm is in circles. This figure shows that the cutoff frequency increases with the decrease of the gate length for MOS transistors. (Modified from Fig. 24 in H. Iwai and H. S. Momose, *Microelectronic Engineering*, vol. 39, no. 1–4 (Dec. 1997), pp. 7–30.)

noise figure tends to become smaller when the MOS transistor is scaled down to smaller dimensions. The interested readers may ask the following question: Can the gate resistance be made smaller by using a small gate width? When the gate width is small, the transconductance g_m becomes small, resulting in a small amplifier gain. According to standard noise theory, the first stage amplifier of a cascade of amplifiers must be a high gain amplifier with a small noise figure F. Thus the gate width cannot be too small.

Since the gate dielectric is an insulating film, the gate input of an MOS transistor is basically capacitive. However, the source of the RF input is basically resistive. The RF input source quite frequently has an internal resistance of 50 Ω. Thus there is some sort of "matching" problem. One solution is to put a 50 Ω resistor across the gate and source of the MOS transistor. However, this 50 Ω resistor will add extra thermal noise, resulting in a poor noise figure. Another possible solution for this problem is to connect the MOS transistor as a common-gate amplifier; the input resistance will be 1/gm. However, Lee[27] pointed out that the noise figure of common-gate amplifier is not good enough. The solution for this problem according to Lee[27] is to add an inductor to the source of the MOS transistor, resulting in a real component (i.e. resistive) of the input admittance of the MOS transistor amplifier; this is known as "inductive source degeneration". (Note: The source here refers to the source electrode of an MOS transistor.) Inductive source degeneration was also discussed in the book by Chang and Sze.[28]

As shown in Fig. 5.12, an inductor L_s can be used to achieve "inductive source degeneration" according to Andreani and Sjoland[29]; however, an inductor L_g is also needed for the purpose of "matching".

5.2.2 *Inductor design*

As discussed above, inductors are required in low noise MOS amplifier for inductive source degeneration. In addition, inductors are also

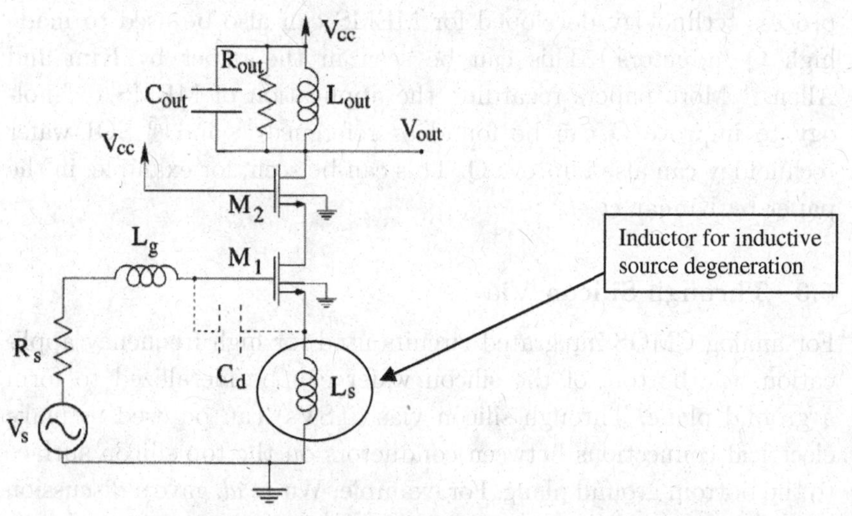

Fig. 5.12 The simplified schematic diagram of a low noise amplifier based on n-channel MOS transistors with inductive source degeneration achieved by the inductor L_s according to Andreani and Sjoland.[29] (Modified from Fig. 1 in P. Andreani and H. Sjoland, *IEEE Transactions on Circuits and Systems II.*, vol. 48, no. 9 (Sep. 2001), pp. 835–841.)

needed in oscillators. In 1974, Greenhouse published a paper on the design of rectangular inductors for microelectronic applications.[30] For many years, the need to increase the Q of inductors used in analog CMOS integrated circuits has been recognized and a large number of papers have been published on how to increase Q. Copper and low-k back-end technology can improve Q. This can be seen in the paper by Huo *et al.*[31]

As discussed by a 2004 review paper on Si wafer technology by Tsuya,[32] high resistivity silicon wafer technology can improve Q of inductors used in silicon based IC technology; in addition, the Q of transmission lines will also be improved. High-resistivity silicon wafers can be easily obtained by silicon wafers manufactured by FZ (float zone) technology. Furthermore, high-resistivity silicon wafers can be easily obtained by silicon wafers manufactured by MCZ (magnetic Czochralski) technology with low oxygen contents. It is also possible to obtain $>1000\,\Omega$cm CZ silicon wafers by using a multi-step annealing technique. MEMS technology can improve Q. (Note: MEMS stands for Micro-Electro-Mechanical-System. It appears that process technology developed for MEMS can also be used to made high Q inductors.) This can be seen in the paper by Kim and Allen.[33] More papers regarding the application of MEMS technology to improve Q can be found as references[34] and.[35] SOI wafer technology can also improve Q. This can be seen, for example, in the paper by Kumar *et al.*[36]

5.3 Through Silicon Via

For analog CMOS integrated circuits used for high frequency application, the bottom of the silicon wafer can be metallized to form a ground plane. Through silicon vias (TSVs) can be used to make electrical connections between conductors on the top silicon surface to the bottom ground plane. For example, Wu *et al.* gave a discussion on the application of TSVs to RFICs.[37,38]

TSVs can also be used for other applications. For example, several silicon wafers can be stacked together and TSVs are used for electrical interconnection. Copper electroplating developed for BEOL technology can also be used for TSV technology.

5.4 MIM Capacitors for Analog CMOS

For analog CMOS integrated circuits, thin film metal-insulator-metal (MIM) capacitors are needed for various purposes. Quite frequently, the insulator thin film is made of a high-k dielectric material. For example, Thomas *et al.* published a paper with the title of "Integration of a high density Ta_2O_5 MIM capacitor following 3D damascene architecture compatible with copper interconnects"; high-k MIM capacitors are integrated into the back end process for analog CMOS applications.[39] For high capacitance per unit area, the high-k dielectric film is "thin"; when there is a voltage across the MIM capacitor, there is a leakage current. As shown in Fig. 5.13, the leakage current versus voltage characteristics of a $TiN/Ta_2O_5/TiN$ MIM capacitor according to Gaillard *et al.* are plotted in a semi-log plot[40];

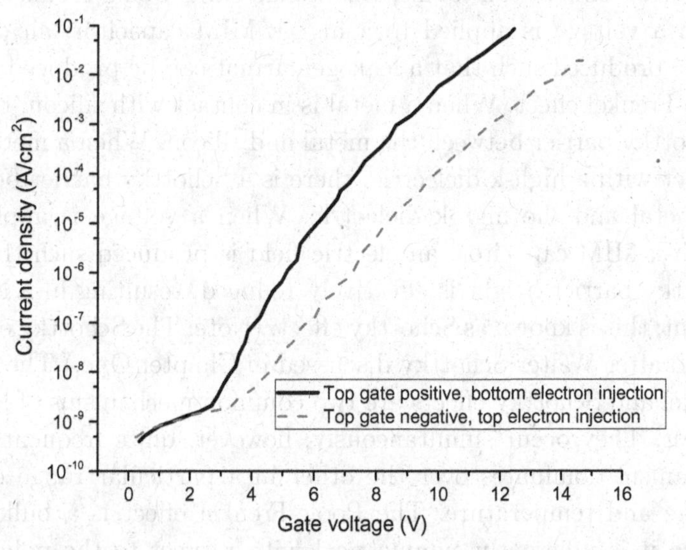

Fig. 5.13 The leakage current versus voltage characteristics of a $TiN/Ta_2O_5/TiN$ MIM capacitor according to Gaillard *et al.*[40] The leakage current due to bottom electron injection when the top TiN electrode was biased positively appeared to be stronger than that due to top electron injection when the top TiN electrode was biased negatively. This could be explained by the observation that the bottom interface was rougher than the top interface. (Modified from Fig. 4 in N. Gaillard, L. Pinzelli, M. Gros-Jean and A. Bsiesy, "In situ electric field simulation in metal/insulator/metal capacitors", *Appl. Phys. Lett.*, vol. 89 (2006), article number 133506.)

Ta_2O_5 was deposited by CVD (chemical vapor deposition). (Note: Thomas *et al.* and Gaillard *et al.* were basically the same group in the company ST Microelectronics, France.) The readers should note that the leakage current versus voltage characteristics in Fig. 5.13 are asymmetrical. The explanation of Gaillard *et al.* was that the bottom interface was rougher than the top interface; however, they did not give a good explanation why the top interface tends to be smoother.

The author (W. S. Lau) has been studying the mechanism of leakage current in high-k MIM capacitors for a couple of years.[41] Silicon is a semiconductor with shallow donors which ionize easily at room temperature. The high-k dielectric can be considered to behave like a large bandgap semiconductor with deep donors which do not ionize easily at room temperature. However, at high electric effect, these deep donors can ionize; this is known as Poole-Frenkel effect. When a voltage is applied to a high-k MIM capacitor, an electric field is produced such that a leakage current can be produced by the Poole-Frenkel effect. When a metal is in contact with silicon, there is a Schottky barrier between the metal and silicon. When a metal is in contact with a high-k dielectric, there is a Schottky barrier between the metal and the high-k dielectric. When a voltage is applied to a high-k MIM capacitor, an electric field is produced such that the Schottky barrier height is effectively reduced, resulting in a leakage current; this is known as Schottky effect. (Note: The Schottky effect is named after Walter Schottky discussed in Chapter One.) The Poole-Frenkel and Schottky effects are two common mechanisms of leakage current. They occur simultaneously; however, quite frequently, one mechanism dominates over the other in a particular range of bias voltage and temperature. The Poole-Frenkel effect is a bulk effect and so it is inherently symmetrical with respect to the polarity of the applied bias voltage. The Schottky effect is an interfacial effect and so it can be asymmetrical with respect to the polarity of the applied bias voltage. It has been experimentally observed that the leakage current versus voltage curve is not symmetrical with respect to the polarity of the applied voltage even though the MIM capacitor seems to be a symmetrical structure. As shown in Fig. 5.13, the leakage current versus voltage curve is not symmetrical with respect to the polarity of the applied voltage even though the MIM capacitor has the same material for the top and bottom electrodes. The author's explanation is that there are two mechanisms involved: (a) asymmetry of roughness and (b) asymmetry of interface defect states density. The theory proposed by the author is that the effective Schottky barrier height between metal and higk-k dielectric can be reduced by "roughness". In addition, the effective Schottky barrier height between metal and higk-k dielectric can also be reduced by the presence of a large quantity of donor-like defect states at the interface between metal and high-k dielectric. (Note: The author's theory is that the deep donors in the bulk of the high-k dielectric may not ionize without the help of a strong electric field but the deep donors at the interface between metal and high-k dielectric may be able to ionize.) These two mechanisms occur simultaneously. The author's experience is that usually Mechanism (a) dominates over Mechanism (b). However, there are some exceptions such that Mechanism (b) dominates over Mechanism (a).

In general, the metal film involved in MIM capacitors is a polycrystalline thin fim. (Note: In general, a polycrystalline film tends to be rougher than an amorphous film.) As shown in Fig. 5.14, the roughness of a polycrystalline metal film is a function of its thickness. Stage (a): When the metal film is very thin, it shows up an "island" structure which is discontinuous, as shown in Fig. 5.15; the film roughness increases with the increase of film thickness. Stage (b): When the metal film becomes thicker, the film roughness decreases with the increase of film thickness because the metal film starts to coalesce resulting in a continuous metal film, as shown in Fig. 5.15. Stage (c): After coalescence, when the metal film becomes thicker, the film roughness increases with the increase of film thickness. For practical MIM capacitors used in microelectronics, the metal films used are in Stage (c).

The author pointed out that CVD (chemical vapor deposition) or ALD (atomic layer deposition) of an amorphous high-k dielectric on a rough metal film has a "surface smoothing" effect such the top interface between the top metal electrode and high-k dielectric is less

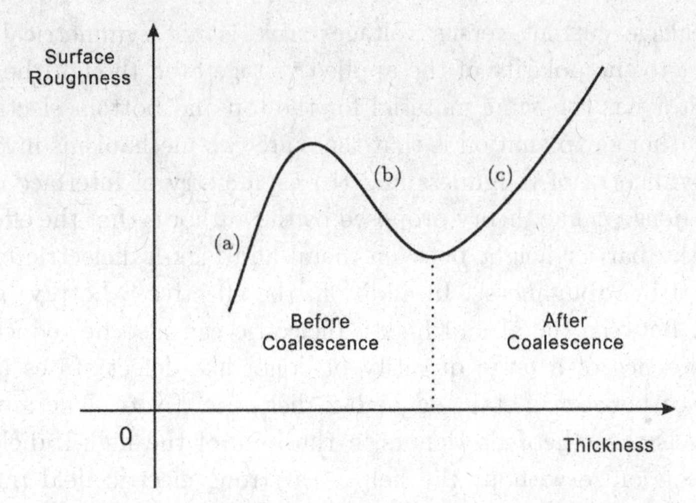

Fig. 5.14 The surface roughness of a polycrystalline metal film. (Note: All practical metal films used in microelectronics are in stage (c). They are polycrystalline and relatively rough.)

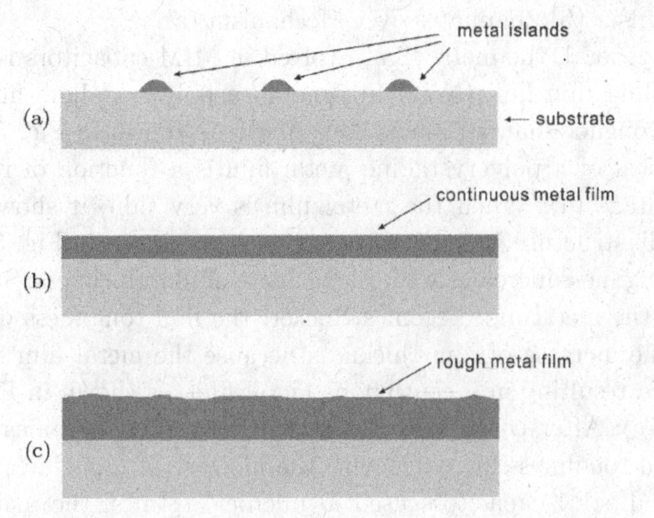

Fig. 5.15 Stage (a): When the metal film is very thin, it shows up an "island" structure which is discontinuous. Stage (b): When the metal film becomes thicker, the metal film starts to coalesce resulting in a continuous metal film. Stage (c): After coalescence, when the metal film becomes thicker, the film roughness increases with the increase of film thickness. For practical MIM capacitors used in microelectronics, the metal films used are in Stage (c).

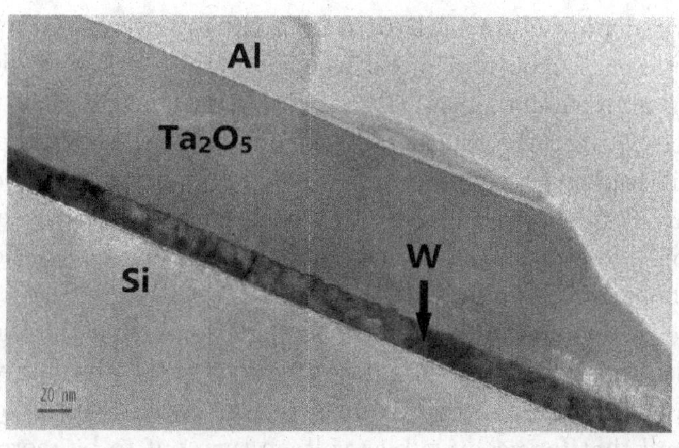

Fig. 5.16 XTEM picture showing that the top interface tends to be smoother than the bottom interface for a Ta_2O_5/W structure. (W thickness is about 20 nm. Ta_2O_5 thickness is about 80 nm.)

rough than the bottom interface between the bottom metal electrode and high-k dielectric.[42,43] As shown in Fig. 5.16, the top interface of CVD Ta_2O_5 tends to be less rough than the bottom interface between the bottom tungsten (W) metal electrode and the high-k dielectric Ta_2O_5. As discussed above, the effective Schottky barrier height between metal and higk-k dielectric can be reduced by "roughness"; this is named as Mechanism (a). According to Gaillard *et al.*,[40] this effect is related to the effective increase of the electric field at sharp points. The author has developed a different opinion from Gaillard *et al.*[40]; the author believes that the effective Schottky barrier height between metal and higk-k dielectric can be influenced by an effect known as the Smoluchowski effect. Metal has free electrons; for a metal surface, the electron cloud tends to protrude slightly beyond the metal surface. In 1941, Smoluchowski pointed out that for a non-flat metal surface, the electron cloud does not follow the surface morphology of metal; in fact, the surface of the electron cloud can be smoother than the metal surface.[44] In this way, the work function of the metal can be modified by the surface morphology of the metal. For example, Li and Li[45] pointed out that the work function of a metal tends to decrease with the increase of the surface roughness of the metal. Thus, MIM capacitors with an amorphous high-k dielectric deposited by CVD or ALD tend to have a bottom interface between the bottom metal electrode and the high-k dielectric showing up a smaller effective Schottky barrier height than the top interface between the top metal electrode and the high-k dielectric. This will lead to an asymmetric I-V characteristics with higher leakage current for bottom electron injection compared to top electron injection. More discussion on the Smoluchowski effect can be found in the book by Venables.[46]

The key evidence for the above explanation is that the logarithm of the leakage current versus the square root of the absolute value of the applied voltage for positive bias voltage applied to the top electrode and that for negative bias voltage turn out to be more or less two parallel lines, as shown in Fig. 5.13 when the magnitude of the applied voltage is bigger than 8 V. This cannot be easily explained by the theory of Gaillard *et al.*.[40] However, this can be more readily explained by the Smoluchowski effect. The leakage current data in Gaillard *et al.*[40] were limited to room temperature data. Leakage current data for more than one temperature can be found in Deloffre *et al.*[47] (Note: Deloffre *et al.* and Gaillard *et al.* were more or less the same research group.) The author analyzed the data from Deloffre *et al.*[47] and found that they can be more readily explained by the Smoluchowski effect. The author used his own $W/Ta_2O_5/W$ samples[48] and found that the leakage current versus bias voltage at various temperatures can be explained by the Smoluchowski effect. For example, the bottom W/Ta_2O_5 interface was rougher and had an effective Schottky barrier height of 0.67 eV whereas the top W/Ta_2O_5 interface was smoother and had an effective Schottky barrier height of 0.88 eV. In addition, the work of Hashimoto *et al.*[49] also supports the above explanation. This can be seen in Fig. 5.17.

As discussed before, the effective Schottky barrier height between metal and higk-k dielectric can also be reduced by the presence of a large quantity of donor-like defect states at the interface between metal and high-k dielectric; this is named as Mechanism (b) above. A high-k dielectric is like a very weakly n-type large bandgap semiconductor with oxygen vacancies serving as deep donors. It can be easily imagined that the oxygen vacancies have a U-shaped

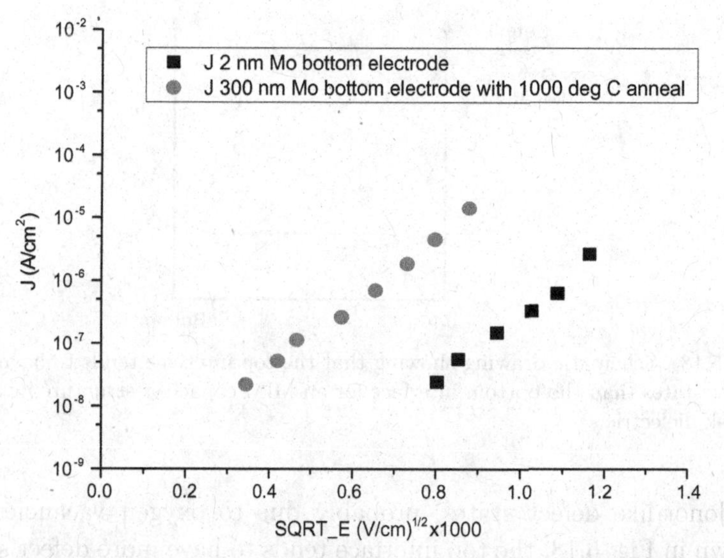

Fig. 5.17 The logarithm of the leakage current density versus the square root of electric field characteristics of $Mo/Ta_2O_5/Mo$ MIM capacitors according to Hashimoto *et al.*[40] The leakage current for a $Mo/Ta_2O_5/Mo$ MIM capacitor with 300 nm thick Mo bottom electrode (with 1000°C annealing before Ta_2O_5 deposition) appeared to be much stronger than that for a $Mo/Ta_2O_5/Mo$ MIM capacitor with 2 nm thick Mo bottom electrode. This could be explained by the much bigger roughness of a 200 nm Mo film compared to a 2 nm Mo film. In addition, the logarithm of the leakage current density versus the square root of electric field characteristics for the 2 MIM capacitors appeared to be perfectly parallel. The author pointed out here that this experimental observation can be explained by the Smoluchowski effect. (Modified from Fig. 4 in C. Hashimoto, H. Oikawa and N. Honma, "Leakage current reduction in thin Ta_2O_5 films for high-density VLSI memories", *IEEE Trans. Electron Dev.*, vol. 36, no. 1 (Jan. 1989), pp. 14–18.)

distribution; there are much more oxygen vacancies at the two interfaces than in the bulk of an MIM capacitor. The author guessed that a high-k dielectric behaves in some way similar to large bandgap semiconductors like gallium nitride (GaN). Hasegawa and Oyama pointed out the effective Schottky barrier height can be reduced by the presence of a large quantity of nitrogen vacancies at the interface according to their "thin surface barrier (TSB)" model.[50,51] In this way, it can be imagined that the effective Schottky barrier height for high-k dielectric can be reduced by the presence of a large quantity

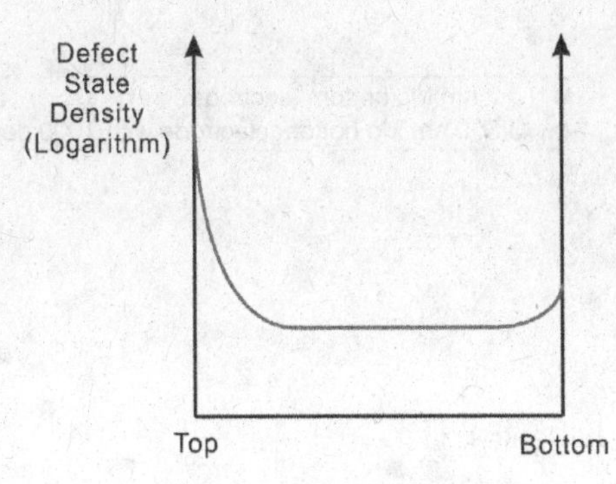

Fig. 5.18 Schematic drawing showing that the top interface tends to have more defect states than the bottom interface for an MIM capacitor structure involving high-k dielectric.

of donor-like defect states, probably due to oxygen vacancies. As shown in Fig. 5.18, the top interface tends to have more defect states than the bottom interface for an MIM capacitor structure involving high-k dielectric. This is because the defect states are related to oxygen vacancies and the bottom tends to be oxidized for a longer time than the top. In addition, some defect states come from the deposition process of the top metal electrode; this also has the effect such that the top interface tends to have more defect states than the bottom interface for an MIM capacitor structure involving high-k dielectric. Mechanism (b) quite frequently produces an effect opposite to Mechanism (a): Mechanism (b) will lead to an asymmetric I-V characteristics with lower leakage current for bottom electron injection compared to top electron injection.

In addition, the work of Lee *et al.*[52] also supports the above explanation. This can be seen in Fig. 5.19.

The logarithm of the leakage current density versus the square root of electric field characteristics for the 2 MIM capacitors appeared to be perfectly parallel. This experimental observation can be explained by the Schottky emission effect.

The leakage current for an Al/$BaTa_2O_6$/ITO MIM capacitor with oxygen plasma treatment (Sample #2) appeared to be much lower than that for an Al/$BaTa_2O_6$/ITO MIM capacitor without oxygen plasma treatment (Sample #1) for negative voltage

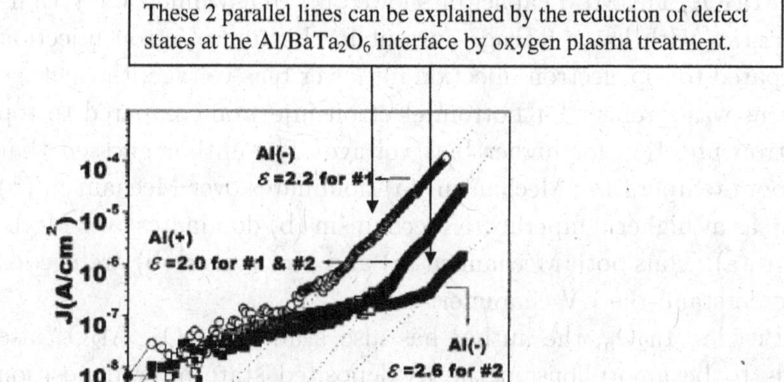

Fig. 5.19 The logarithm of the leakage current density versus the square root of electric field characteristics of $Al/BaTa_2O_6/ITO$ MIM capacitors according to Lee *et al.*[52] Sample #1 did not have an oxygen plasma treatment. Sample #2 had an oxygen plasma treatment.

applied to the top Al metal electrode. This can be explained by the much lower defect state density at the $Al/BaTa_2O_6$ interface for the $Al/BaTa_2O_6/ITO$ MIM capacitor with oxygen plasma treatment (Sample #2) compared to the $Al/BaTa_2O_6/ITO$ MIM capacitor without oxygen plasma treatment (Sample #1), resulting in a larger effective Schottky barrier height for Sample #2 compared to Sample #1. (Note: ITO is indium tin oxide which is a transparent conductor.) (Modified from Fig. 5 in Y.-H. Lee, Y.-K. Kim, D.-H. Kim, B.-K. Ju and M.-H. Oh, "Conduction mechanisms in barium tantalates films and modification of interfacial barrier height", *IEEE Trans. Electron Dev.*, vol. 47, no. 1 (Jan. 2000), pp. 71–76.)

In 2003, Sun *et al.*[53] reported their experimental results on $TiN/Ta_2O_5/TiN$ MIM capacitors. At room temperature (27°C for Sun *et al.*), the leakage current versus voltage characteristics of their $TiN/Ta_2O_5/TiN$ MIM capacitor showed up an asymmetric I-V characteristics with higher leakage current for bottom electron injection compared to top electron injection. However, at higher temperature (125°C), the leakage current versus voltage characteristics of their

TiN/Ta_2O_5/TiN MIM capacitor showed up an asymmetric I-V characteristics with lower leakage current for bottom electron injection compared to top electron injection for lower bias voltage; the leakage current was stronger for bottom electron injection compared to top electron injection for higher bias voltage. The author guessed that at room temperature Mechanism (a) dominates over Mechanism (b) whereas at higher temperature Mechanism (b) dominates over Mechanism (a). Thus both Mechanism (a) and Mechanism (b) are needed to understand the I-V characteristics.

Besides Ta_2O_5, the author has also studied Al_2O_3. Al_2O_3 also tends to be amorphous in the as deposited state after deposition by ALD. Previously, the author pointed out that Al_2O_3 deposited by ALD has a surface smoothing effect. In this book, the author concentrates on high-k dielectric materials like Ta_2O_5 or Al_2O_3 which tends to be amorphous in the as-deposited state after CVD or ALD. The readers should note that other high-k dielectric materials like titanium oxide (TiO_2) or zirconium oxide (ZrO_2) or hafnium oxide (HfO_2) tend to behave in a different way in the as-deposited state after CVD or ALD. This will be discussed later.

In this book, the author would also like to point out to the readers that the "surface smoothing" effect of CVD or ALD sometimes may not difficult to observe by atomic force microscopy (AFM). Experimentally, this effect is sometimes observed by AFM; however, sometimes an apparently contradictory effect was observed.[54] There exist some extra factors such that it has been difficult to observe the "surface smoothing effect" of CVD or ALD by AFM. The readers can imagine that an experiment can be done in the following manner. A rough polycrystalline metal film is deposited by sputtering at room temperature on a smooth Si substrate. AFM is performed to measure the RMS surface roughness, resulting in a value of SR1. Then Al_2O_3 is deposited by ALD at 300°C, which is significantly higher than room temperature. AFM is performed again to measure the RMS surface roughness, resulting in a value of SR2. If there is a "surface smoothing effect", it is expected that SR2 is smaller than SR1. However, in reality, this may not be experimentally observed; it is possible that SR2 is larger than SR1 according to experimental observation.

There are some additional factors which can affect the roughness of the metal/insulator interfacial region, for example: (1) non-uniform oxidation of the bottom metal electrode during the deposition of the insulator and (2) the grain growth of the bottom metal electrode. Thus after ALD, the actual RMS surface roughness of the metal/insulator interfacial region can be SR* such that SR* > SR1. SR2 may be smaller than SR* because of the "surface smoothing effect" but SR* is difficult to measure and the experimentalist may only be able to see SR2 > SR1 instead of SR2 < SR1. In this way, the confused experimentalist may not be able to see the "surface smoothing effect" of ALD. The above theory is explained more clearly in Fig. 5.20; for example, SR1 = 0.6 nm while SR* = 1 nm; SR2 = 0.7 nm. Comparing SR2 with SR*, the surface smoothing effect of ALD can be seen. However, in reality, SR* cannot be measured unless the ALD Al_2O_3 layer is removed. Comparing SR2 with SR1, the experimentalist may wrongly conclude that there is no surface smoothing

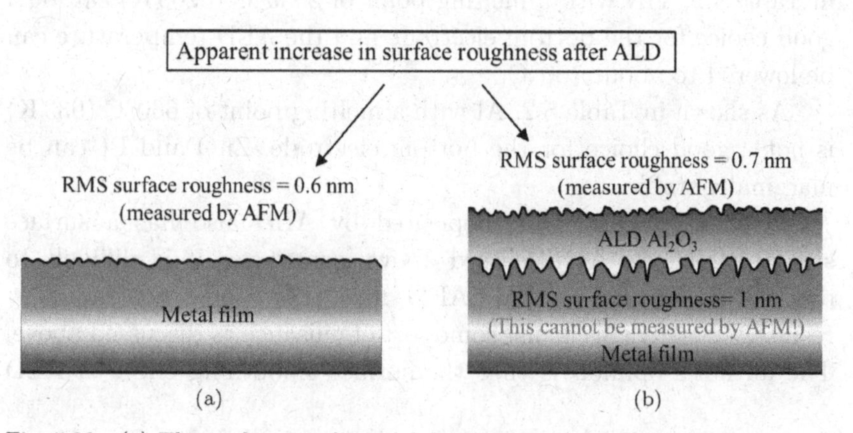

Fig. 5.20 (a) The surface roughness of the top of a metal film can be measured by AFM, resulting in SR1. (b) After the deposition of an Al_2O_3 film by ALD, the surface roughness of the top of the Al_2O_3 film can be measured, resulting in SR2. However, the surface roughness of the metal film may change during the ALD process from SR1 to SR*, which cannot be measured by AFM. For example, SR1 = 0.6 nm while SR* = 1 nm; SR2 = 0.7 nm. Comparing SR2 with SR*, the surface smoothing effect of ALD can be seen. However, in reality, SR* cannot be measured. Comparing SR2 with SR1, the experimentalist may wrongly conclude that there is no surface smoothing effect by ALD because the surface roughness apparently increases instead of decreases.

effect by ALD because the surface roughness apparently increases instead of decreases. The author has tried to measure SR* by using a selective etch to etch away the ALD Al_2O_3 layer and then perform atomic force microcopy on the top of the underlying metal film after selective etching.[55]

To tackle the first problem mentioned in the previous paragraph, a relatively chemically inert metal can be considered as the material for the bottom electrode. To tackle the second problem that a metal film can be slightly unstable even at a relatively low temperature, a conductor which is stable when heated should be considered as the material for the bottom electrode. It can be easily imagined that the bulk, grain boundary and surface self-diffusivities for a material at given temperature, tend to be lower the higher the melting temperature of that material. Thus it can be predicted that a material with a high melting point tends to be more stable when heated. In addition, the temperature of ALD used should be smaller. As shown in Table 5.2, TiN with a melting point of 2930°C (3203 K) can be a good choice for the bottom electrode and the ALD temperature can be lowered to about 150°C.

As shown in Table 5.2, Al with a melting point of 660°C (933 K) is not a good choice for the bottom electrode. ZnO and Pt can be marginally stable.

Besides Ta_2O_5, Al_2O_3 deposited by ALD also has a surface smoothing effect. As discussed above, sometimes it is difficult to detect this effect by AFM. AFM may show some contradictory results. Actually this is just some sort of nuisance as discussed above. The author's opinion is that the surface smoothing effect of ALD

Table 5.2. Melting points of various materials.

Material	Melting point (K)
TiN	3203
ZnO	2248
Pt	2041
Al	933

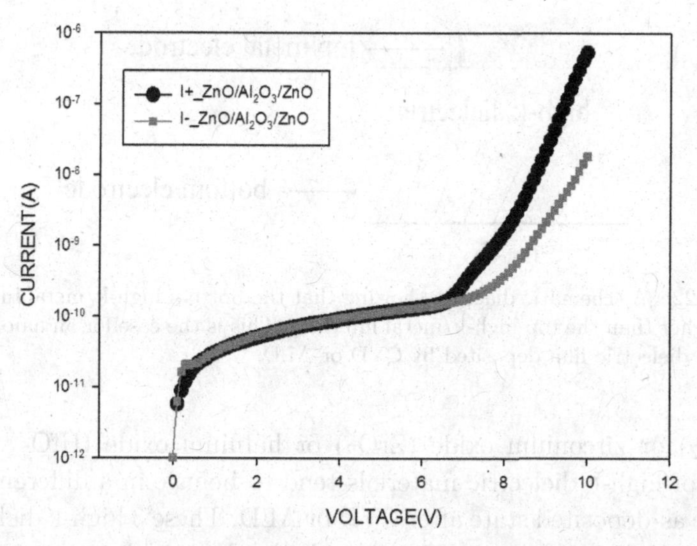

Fig. 5.21 The logarithm of the leakage current versus the voltage characteristics of $ZnO/Al_2O_3/ZnO$ MIM capacitors according to Lau *et al.*[54] ZnO deposited by ALD is conductive and can be used as an electrode. The thickness of the Al_2O_3 deposited by ALD is about 20 nm. The leakage current is stronger for positive voltage applied to the top ZnO electrode than that for negative voltage applied to the top ZnO electrode. The author pointed out here that this experimental observation can be explained by the Smoluchowski effect; the bottom Al_2O_3/ZnO interface is rougher than the top Al_2O_3/ZnO interface. (Modified from Fig. 6 in Lau *et al.*[54])

Al2O3 exists beyond doubt and can be used to interpret the leakage current versus voltage characteristics. As shown in Fig. 5.21, the leakage current is stronger for positive voltage applied to the top ZnO electrode than that for negative voltage applied to the top ZnO electrode. ZnO is polycrystalline and the surface is slightly rough.[54] Al_2O_3 deposited by ALD has a surface smoothing effect such that the bottom Al_2O_3/ZnO interface is rougher than the top Al_2O_3/ZnO interface. A schematic diagram showing that the bottom high-k/metal interface is rougher than the top high-k/metal interface is given in this book as Fig. 5.22.

As discussed above, besides the class of high-k dielectric materials like tantalum oxide (Ta_2O_5) and aluminum oxide (Al_2O_3), there is another class of high-k dielectric materials like titanium oxide

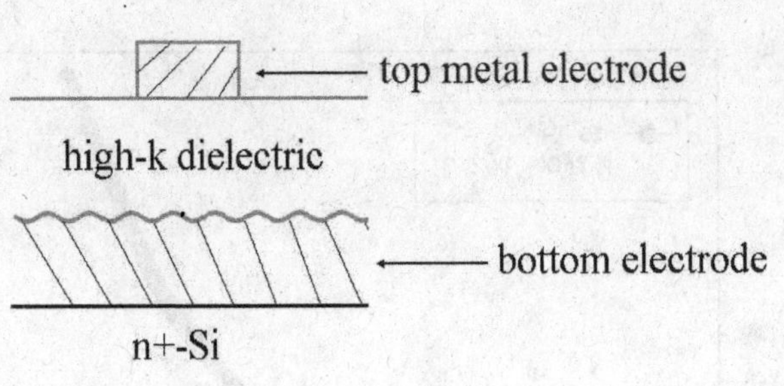

Fig. 5.22 A schematic diagram showing that the bottom high-k/metal interface is rougher than the top high-k/metal interface. This is the case for an amorphous high-k dielectric film deposited by CVD or ALD.

(TiO_2) or zirconium oxide (ZrO_2) or hafnium oxide (HfO_2). This class of high-k dielectric materials tend to behave in a different way in the as-deposited state after CVD or ALD. These 3 high-k dielectric material tend to be amorphous in the as deposited state when the thickness is very small, for example, a few nanometers. For example, Kim *et al.*[56] pointed out that as deposited ALD TiO_2 was amorphous, when the thickness was very small, whereas as deposited ALD TiO_2 was polycrystalline, when the thickness was large. Similarly, Baryshnikova *et al.*[57] pointed out that as deposited CVD TiO_2 was amorphous, when the thickness was very small, whereas as deposited CVD TiO_2 was polycrystalline, when the thickness was large. Furthermore, Weinreich *et al.*[58] pointed out that as deposited ALD ZrO_2 was amorphous, when the thickness was very small, whereas as deposited ALD ZrO_2 was polycrystalline, when the thickness was large. The same is true for ALD HfO2 as explained by Nguyen *et al.*[59] and Seo *et al.*[60]

The "surface smoothing" effect of CVD or ALD can still be applicable to the class of high-k dielectric materials like hafnium oxide (HfO_2) when the thickness is very small, for example, a few nanometers. There exists a critical thickness such that the film deposited by ALD with thickness bigger than the critical thickness is polycrystalline. In 2015, Nie *et al.* managed to give an explanation of this phenomenon based on thermodynamics.[61] For an MIM capacitor with the same electrode material for the top electrode and the bottom electrode, when the thickness is below the critical thickness,

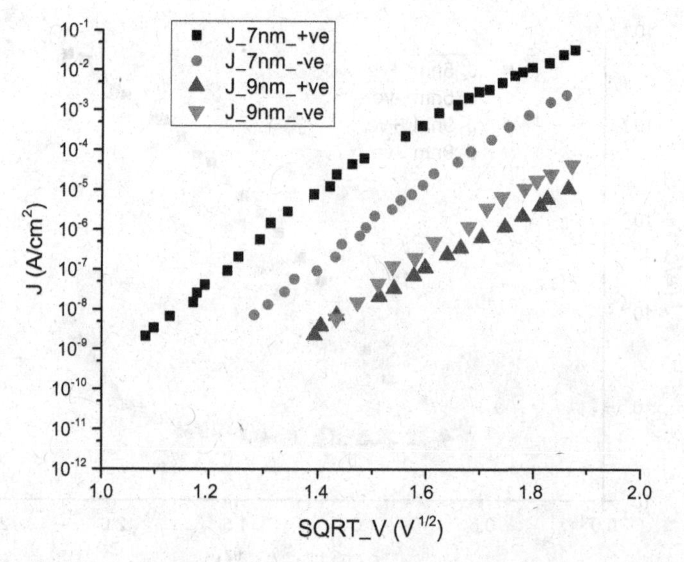

Fig. 5.23 For 7 nm amorphous ALD $Zr_{(1-x)}Al_xO_2$, the leakage current density is stronger for positive voltage applied to top metal electrode than for negative voltage applied to top metal electrode. For 9 nm polycrystalline ALD $Zr_{(1-x)}Al_xO_2$, the leakage current density is stronger for negative voltage applied to top metal electrode than for positive voltage applied to top metal electrode. (Modified from Fig. 1 in W. Weinreich, R. Reiche, M. Lemberger, G. Jegert, J. Muller, L. Wilde, S. Teichert, J. Heitmann, E. Erben, L. Oberbeck, U. Schroder, A. J. Bauer and H. Ryssel, "Impact of interface variations on J-V and C-V polarity asymmetry of MIM capacitors with amorphous and crystalline $Zr_{(1-x)}Al_xO_2$ films", *Microelectronic Engineering*, vol. 86 (2009), pp. 1826–1829.)

the leakage current is stronger for negative voltage applied to the bottom electrode. When the thickness is above the critical thickness, the leakage current is stronger for negative voltage applied to the top electrode. As shown in Fig. 5.23, this effect can be seen in ZrAlO.[62] As shown in Fig. 5.24, this effect can be seen in HfO_2.[63] As shown in Fig. 5.25, there is a corresponding sharp change in the surface roughness of HfO_2 grown by ALD on Ru.[63]

The above discussion is for a situation when there is a significant difference between the roughness of the top metal/high-k interface and the bottom metal/high-k interface. There exists a situation when there is no significant difference between the roughness of the top metal/high-k interface and the bottom metal/high-k interface. For

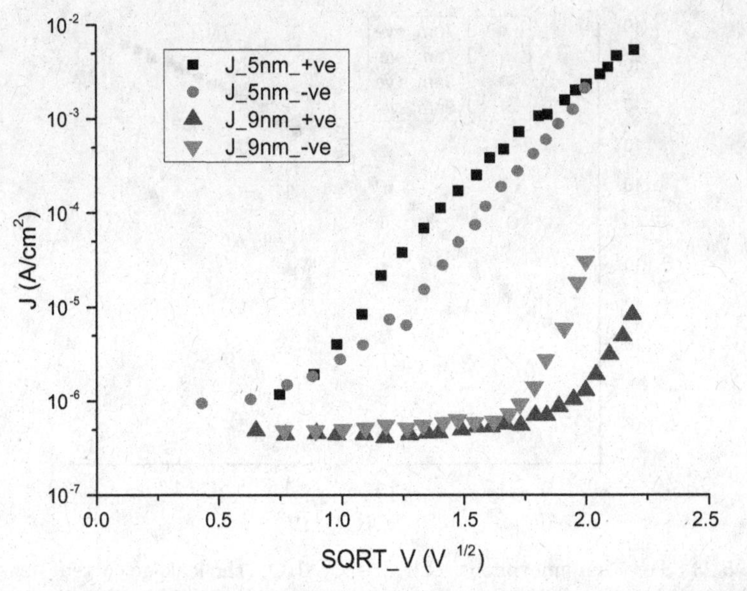

Fig. 5.24 For 5 nm amorphous ALD HfO_2, the leakage current density is stronger for positive voltage applied to top metal electrode than for negative voltage applied to top metal electrode. For 9 nm polycrystalline ALD HfO_2, the leakage current density is stronger for negative voltage applied to top metal electrode than for positive voltage applied to top metal electrode. (Modified from Fig. 3 in J.-H. Kim, S.-G. Yoon, S.-J. Yeom, H.-K. Woo, D.-S. Kil, J.-S. Roh and H.-C. Sohn, "Electrical properties in high-k HfO_2 capacitors with an equivalent oxide thickness of 9 Å on Ru metal electrode", *Electrochemical and Solid-State Letters*, vol. 8, no. 6 (2005), pp. F17–F19.) (*Note*: Kim *et al.* did not state clearly the polarity of applied voltage in their paper. The author used his theoretical understand to deduce that positive voltage applied to top metal electrode in the above figure corresponds to negative voltage in the figures of Kim *et al.*)

example, when the bottom metal is very thin, the bottom metal is smooth; the top metal/high-k interface and the bottom metal/high-k interface will be smooth after ALD of Al_2O_3. However, the top interface may have more defect states than the bottom interface such that the effective Schottky barrier height is smaller for the top interface than that of the bottom interface. This may be the situation for Meng *et al.*[64] In addition, when the bottom metal is amorphous, the bottom metal is smooth; the top metal/high-k interface and the bottom metal/high-k interface will be smooth after ALD of Al_2O_3.

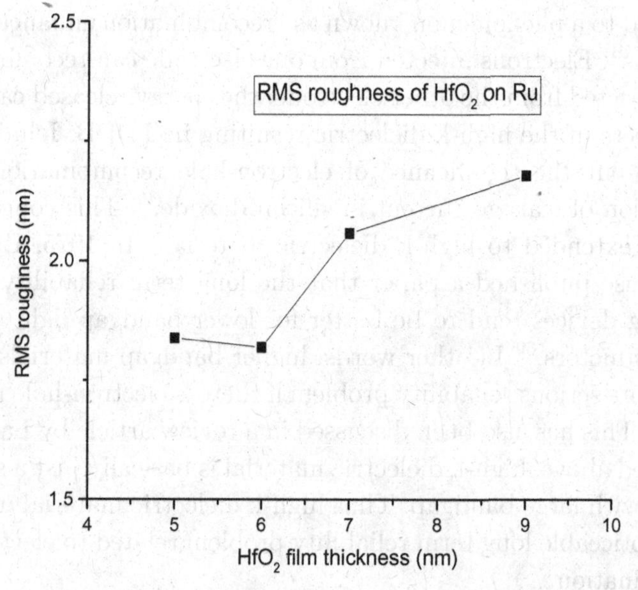

Fig. 5.25 The variation of surface roughness of ALD HfO_2 grown on Ru as a function of the ALD HfO_2 film thickness. (Modified from Fig. 1 in J.-H. Kim, S.-G. Yoon, S.-J. Yeom, H.-K. Woo, D.-S. Kil, J.-S. Roh and H.-C. Sohn, "Electrical properties in high-k HfO_2 capacitors with an equivalent oxide thickness of 9 Å on Ru metal electrode", *Electrochemical and Solid-State Letters*, vol. 8, no. 6 (2005), pp. F17–F19.)

However, the top interface may have more defect states than the bottom interface such that the effective Schottky barrier height is smaller for the top interface than that of the bottom interface. This may be the situation for Cowell *et al.*[65] when they used amorphous TiAl for both top and bottom electrodes and ALD Al_2O_3 as high-k dielectric.

Regarding the long term reliability of high-k MIM capacitors, TDDB (time dependent dielectric breakdown) can be a tissue. Sato *et al.*[66] discussed this problem and pointed out the importance of electron injection. Hirano *et al.*[67] discussed this problem and pointed out the importance of both electron injection and hole injection. Judging from a simple principle of symmetry, the author believes that both electron injection and hole injection can be important. The author guessed that TDDB of high-k dielectric MIM capacitors is related to a phenomenon known as "recombination enhanced defect reactions". Electrons injected from one electrode can recombine with holes injected from the other electrode; the energy released can generate defects in the high-k dielectric resulting in TDDB. Ielmini *et al.* pointed out the significance of electron-hole recombination in the generation of leakage current in silicon dioxide.[68] This concept may also be extended to high-k dielectric materials. In 1975, Ettenberg and Neuse published a paper that the long term reliability of light emitting devices tend to be better for lower bandgap light emitting semiconductors.[69] In other words, higher bandgap materials tend to have more serious reliability problem if there is electron-hole recombination. This has also been discussed in a review article by Lang.[70] As discussed above, high-k dielectric material is basically just a semiconductor with large bandgap. Thus high-k dielectric material may have some noticeable long term reliability problem related to electron-hole recombination.

5.5 Mixed Signal CMOS

A discussion of the application of analog CMOS technology for wireless systems was given by Matsuzawa in 2006.[71] In 2008, Matsuzawa published a paper on the application of both digital and analog CMOS technology for RF.[72] The combination of both digital and analog CMOS technology is known as mixed-signal CMOS technology. For example, a book on CMOS mixed-signal circuit design was published by Jacob Baker.[73]

References

[1] M. R. Haskard and I. C. May, *Analog VLSI Design: nMOS and CMOS*, Prentice Hall, Englewood Cliffs, New Jersey, USA, 1988, pp. 1–243.

[2] P. R. Gray, P. J. Hurst, S. H. Lewis and R. G. Meyer, *Analysis and Design of Analog Integrated Circuits*, 5th edition, Wiley, New York, 2009, pp. 1–881.

[3] J. M. Early, "Effects of space-charge layer widening in junction transistors", *Proc. IRE*, vol. 40, no. 11 (Nov. 1952), pp. 1401–1406.

[4] V. Kilchytska, A. Neve, L. Vancaillie, D. Levacq, S. Adriaensen, H. van Meer, K. De Meyer, C. Raynaud, M. Dehan, J.-P. Raskin and D. Flandre, "Influence of device engineering on the analog and RF performances of

SOI MOSFETs", *IEEE Trans. Electron Dev.*, vol. 50, no. 3 (Mar. 2003), pp. 577–588.
[5] M.-C. Chang, C.-S. Chang, C.-P. Chao, K.-I. Goto, M. Ieong, L.-C. Lu and C. H. Diaz, "Transistor- and circuit-design optimization for low-power CMOS", *IEEE Trans. Electron Dev.*, vol. 55, no. 1 (Jan. 2008), pp. 84–95.
[6] J.-C. Guo, "Halo and LDD engineering for multiple Vth high performance analog CMOS devices", *IEEE Trans. Semiconductor Manufacturing*, vol. 20, no. 3 (Aug. 2007), pp. 313–322.
[7] M. Miyamoto, K. Toyota, K. Seki and T. Nagano "Asymmetrically-doped buried layer (ADB) structure CMOS for low-voltage mixed analog-digital applications", *Symposium on VLSI Technology Digest of Technical Papers*, pp. 102–103, 1996.
[8] A. Chatterjee, K. Vasanth, D. T. Grider, M. Nandakumar, G. Pollack, R. Aggarwal, M. Rodder and H. Shichijo, "Transistor design issues in integrating analog functions with high performance digital CMOS", *Symposium on VLSI Technology Digest of Technical Papers*, pp. 147–148 (1999).
[9] A. S. Roy, S. P. Mudanai and M. Stettler, "Mechanism of long-channel drain-induced barrier lowering in halo MOSFETs", *IEEE Trans. Electron Dev.*, vol. 58, no. 4 (Apr. 2011), pp. 979–984.
[10] K. M. Cao, W. Liu, X. Jin, K. Vashanth, K. Green, J. Krick, T. Vrotsos and C. Hu, "Modeling of pocket implanted MOSFETs for anomalous analog behavior", *IEDM Tech. Dig.*, pp. 171–174 (1999).
[11] D. Buss, "Device issues in the integration of analog/RF functions in deep submicron digital CMOS," *IEDM Tech. Dig.*, pp. 423–426 (1999).
[12] P. A. Stolk, H. P. Tuinhout, R. Duffy, E. Augendre, L. P. Bellefroid, M. J. B. Bolt, J. Croon, C. J. J. Dachs, F. R. J. Huisman, A. J. Moonen, Y. V. Ponomarev, R. F. M. Roes, M. Da Rold, E. Seevinck, K. N. Sreerambhatla, R. Surdeanu, R. M. D. A. Velghe, M. Vertregt, M. N. Webster, N. K. J. van Winkelhoff, and A. T. A. Zegers-Van Duijnhoven, "CMOS device optimization for mixed-signal technologies," *IEDM Tech. Dig.*, pp. 215–218 (2001).
[13] V. Subramanian, B. Parvais, J. Borremans, A. Mercha, D. Linten, P. Wambacq, J. Loo, M. Dehan, C. Gustin, N. Collaert, S. Kubicek, R. Lander, J. Hooker, F. Cubaynes, S. Donnay, M. Jurczak, G. Groeseneken, W. Sansen and S. Decoutere, "Planar bulk MOSFETs versus FinFETs: An analog/RF perspective", *IEEE Trans. Electron Dev.*, vol. 53, no. 12 (Dec. 2006), pp. 3071–3079.
[14] L. L. Lewyn, T. Ytterdal, C. Wulff and K. Martin, "Analog circuit design in nanoscale CMOS technologies", *Proc. IEEE*, vol. 97, no. 10 (Oct. 2009), pp. 1687–1714.
[15] J. A. Croon, W. Sansen and H. E. Maes, *Matching Properties of Deep Submicron MOS Transistors*, Springer, P. O. Box 17, 3300 AA Dordrecht, The Netherlands, 2005, pp. 1–206.
[16] A. Marshall, *Mismatch and Noise in Modern IC Processes*, Morgan & Claypool Publishers, 2009, pp. 1–139.
[17] K. R. Lakshmikumar, R. A. Hadaway and M. A. Copeland, "Characterization and modeling of mismatch in MOS transistors for precision

analog design", *IEEE J. Solid-State Circuits*, vol. SC-21, no. 6 (Dec. 1986), pp. 1057–1066.

[18] C. S. G. Conroy, W. A. Lane and M. A. Moran, "A comment on 'Characterization and modeling of mismatch in MOS transistors for precision analog design'", *IEEE J. Solid-State Circuits*, vol. 23, no. 1 (Feb. 1988), pp. 294–296.

[19] M. J. M. Pelgrom, H. P. Tuinhout and M. Vertregt, "Transistor matching in analog CMOS applications," *IEDM Tech. Dig.*, pp. 915–918 (1998).

[20] M. J. M. Pelgrom, A. C. J. Duinmaijer and A. P. G. Welbers, "Matching properties of MOS transistors", *IEEE J. Solid-State Circuits*, vol. 24, no. 5 (Oct. 1989), pp. 1433–1440.

[21] X. Yuan, T. Shimizu, U. Mahalingam, J. S. Brown, K. Z. Habib, D. G. Tekleab, T.-C. Su, S. Satadru, C. M. Olsen, H. Lee, L.-H. Pan, T. B. Hook, J.-P. Han, J.-E. Park, M.-H. Na and K. Rim, "Transistor mismatch properties in deep-submicrometer CMOS technologies", *IEEE Trans. Electron Dev.*, vol. 58, no. 2 (Feb. 2011), pp. 335–342.

[22] U. Schaper and J. Einfeld, "Matching model for planar bulk transistors with halo implantation", *IEEE Electron Dev. Lett.*, vol. 32, no. 7 (July 2011), pp. 859–861.

[23] T. Merelle, G. Curatola, A. Nackaerts, N. Collaert, M. J. H. van Dal, G. Doornbos, T. S. Doorn, P. Christie, G. Vellianitis, B. Duriez, R. Duffy, B. J. Pawlak, F. C. Voogt, R. Rooyackers, L. Witters, M. Jurczak and R. J. P. Lander, "First observation of FinFET specific mismatch behavior and optimization guidelines for SRAM scaling", *IEDM Tech. Dig.*, pp. 1–4 (2008).

[24] D. K. Shaeffer and T. H. Lee, "A 1.5-V, 1.5-GHz CMOS low noise amplifier", *IEEE J. Solid-State Circuits*, vol. 32, no. 5 (May 1997), pp. 745–759.

[25] D. K. Shaeffer and T. H. Lee, "Corrections to A 1.5-V, 1.5-GHz CMOS low noise amplifier", *IEEE J. Solid-State Circuits*, vol. 40, no. 6 (June 2005), pp. 1397–1398.

[26] H. Iwai and H. S. Momose, "Technology towards low power/low voltage and scaling of MOSFETs", *Microelectronic Engineering*, vol. 39, no. 1–4 (Dec. 1997), pp. 7–30.

[27] T. H. Lee, *The Design of CMOS Radio-frequency Integrated Circuits*, Cambridge University Press, Cambridge, UK, 1998, pp. 1–486.

[28] C. Y. Chang and S. M. Sze, *ULSI Devices*, John Wiley & Sons, New York, USA, 2000, p. 505.

[29] P. Andreani and H. Sjoland, "Noise optimization of an inductively degenerated CMOS low noise amplifier", *IEEE Transactions on Circuits and Systems II.*, vol. 48, no. 9 (Sep. 2001), pp. 835–841.

[30] H. M. Greenhouse, "Design of planar rectangular microelectronic inductors", *IEEE Transactions on Parts, Hybrids and Packaging*, vol. PHP-10, no. 2 (June 1974), pp. 101–109.

[31] X. Huo, K. J. Chen and P. C. H. Chan, "Silicon-based high-Q inductors incorporating electroplated copper and low-k BCB dielectric", *IEEE Electron Dev. Lett.*, vol. 23, no. 9 (Sep. 2002), pp. 520–522.

[32] H. Tsuya, "Present status and prospect of Si wafers for Ultra Large Scale Integration", *Jpn. J. Appl. Phys.*, vol. 43, no. 7A (July 2004), pp. 4055–4067.

[33] Y.-J. Kim and M. G. Allen, "Surface micromachined solenoid inductors for high frequency applications", *IEEE Transactions on Components, Packaging, and Manufacturing Technology Part C*, vol. 21, no. 1 (Jan. 1998), pp. 26–33.

[34] I. Zine-El-Abidine and M. Okoniewski, "CMOS-compatible micromachined toroid and solenoid inductors with high Q-factors", *IEEE Electron Dev. Lett.*, vol. 28, no. 3 (Mar. 2007), pp. 226–228.

[35] J. C. Wu and M. E. Zaghloul, "CMOS micromachined inductors with structure supports for RF mixer matching networks", *IEEE Electron Dev. Lett.*, vol. 29, no. 11 (Nov. 2008), pp. 1209–1211.

[36] M. Kumar, Y. Tan and J. K.O. Sin, "Excellent cross-talk isolation, high-Q inductors and reduced self-heating in a TFSOI technology for system-on-a-chip applications", *IEEE Trans. Electron Dev.*, vol. 49, no. 4 (Apr. 2002), pp. 584–589.

[37] J. H. Wu, J. Scholvin and J. A. del Alamo, "A through-wafer interconnect in silicon for RFICs", *IEEE Trans. Electron Dev.*, vol. 51, no. 11 (Nov. 2004), pp. 1765–1771.

[38] J. H. Wu and J. A. del Alamo, "Fabrication and characterization of through-substrate interconnects", *IEEE Trans. Electron Dev.*, vol. 57, no. 6 (June 2010), pp. 1261–1268.

[39] M. Thomas, A. Farcy, N. Gaillard, C. Perrot, M. Gros-Jean, I. Matko, M. Cordeau, W. Saikaly, M. Proust, P. Caubet, E. Deloffre, S. Cremer, S. Bruyere, B. Chenevier and J. Torres, "Integration of a high density Ta_2O_5 MIM capacitor following 3D damascene architecture compatible with copper interconnects", *Microelectronic Engineering*, vol. 83, no. 11–12 (November–December 2006), pp. 2163–2168.

[40] N. Gaillard, L. Pinzelli, M. Gros-Jean and A. Bsiesy, "In situ electric field simulation in metal/insulator/metal capacitors," *Appl. Phys. Lett.*, vol. 89 (2006), article number 133506.

[41] W. S. Lau, "An extended unified Schottky-Poole-Frenkel theory to explain the current-voltage characteristics of thin film metal-insulator-metal capacitors with examples for various high-k dielectric materials", *ECS J. Solid State Sci. Technol.*, vol. 1, no. 6 (2012), pp. N139–N148.

[42] W. S. Lau, J. Zhang, X. Wan, J. K. Luo, Y. Xu and H. Wong, "Surface smoothing effect of an amorphous thin film deposited by atomic layer deposition on a surface with nano-sized roughness," *AIP Advances*, vol. 4 (2014) article no. 027120.

[43] W. S. Lau, J. Zhang, X. Wan, H. Wong, J. K. Luo and Y. Xu, "Surface smoothing effect of an amorphous thin film deposited by chemical vapor deposition or atomic layer deposition," *ECS Trans.*, vol. 60(1) (2014), pp. 527–531.

[44] R. Smoluchowski, "Anisotropy of the electronic work function of metals," *Phys. Rev.*, vol. 60 (1941), pp. 661–674.

[45] W. Li and D. Y. Li, "On the correlation between surface roughness and work function in copper", *J. Chem. Phys.*, vol. 122, article number 064708, 2005.

[46] J. A. Venables, *Introduction to Surface and Thin Film Processes*, Cambridge University Press, Cambridge, 2003, p. 195.

[47] E. Deloffre, L. Montes, G. Ghibaudo, S. Bruyere, S. Blonkowski, S. Becu, M. Gros-Jean and S. Cremer, "Electrical properties in low temperature range (5 K–300 K) of tantalum oxide dielectric MIM capacitors", *Microelectronics Reliability*, vol. 45 (2005), pp. 925–928.

[48] D. Q. Yu, W. S. Lau, H. Wong, X. Feng, S. Dong and K. L. Pey, "The variation of the leakage current characteristics of $W/Ta_2O_5/W$ MIM capacitors with the thickness of the bottom W electrode", *Microelectronics Reliability*, vol. 61 (2016), pp. 95–98.

[49] C. Hashimoto, H. Oikawa and N. Honma, "Leakage current reduction in thin Ta_2O_5 films for high-density VLSI memories", *IEEE Trans. Electron Dev.*, vol. 36, no. 1 (Jan. 1989), pp. 14–18.

[50] H. Hasegawa and S. Oyama, "Mechanism of anomalous current transport in n-type GaN Schottky contacts," *J. Vac. Sci. Technol. B*, vol. 20, no. 4 (July/August 2002), pp. 1647–1655.

[51] T. Hashizume, J. Kotani and H. Hasegawa, "Leakage mechanism in GaN and AlGaN Schottky interfaces," *Appl. Phys. Lett.*, vol. 84, no. 24 (14 June 2004), pp. 4884–4886.

[52] Y.-H. Lee, Y.-K. Kim, D.-H. Kim, B.-K. Ju and M.-H. Oh, "Conduction mechanisms in barium tantalates films and modification of interfacial barrier height", *IEEE Trans. Electron Dev.*, vol. 47, no. 1 (Jan. 2000), pp. 71–76.

[53] H. Sun, K. M. Lau, E. Aksen and N. Bell, "Improvement of tantalum pentoxide metal-insulator-metal capacitors for SiGe BiCMOS technology," *MRS Symp. Proc.*, vol. 783 (2004), pp. B.7.6.1–B.7.6.6.

[54] W. S. Lau, D. Q. Yu, X. Wang, H. Wong and Y. Xu, "Mechanism of I-V symmetry of MIM capacitors based on high-k dielectric," *Proceedings of CSTIC 2015 (China Semiconductor Technology International Conference*, Shanghai, 2015, IEEE), pp. 1–3, 2015.

[55] W. S. Lau, D. Q. Yu, X. Wang, H. Wong and Y. Xu, "Confirmation of the surface smoothing effect of atomic layer deposition and the physical mechanism responsible for such an effect," *Proceedings of CSTIC 2016 (China Semiconductor Technology International Conference*, Shanghai, 2016, IEEE), pp. 1–3, 2016.

[56] W. D. Kim, G. W. Hwang, O. S. Kwon, S. K. Kim, M. Cho, D. S. Jeong, S. W. Lee, M. H. Seo, C. S. Hwang, Y.-S. Min and Y. J. Cho, "Growth characteristics of atomic layer deposited TiO_2 thin films on Ru and Si electrodes for memory capacitor applications", *J. Electrochem. Soc.*, vol. 152, no. 8 (2005), pp. C552–C559.

[57] M. Baryshnikova, L. Filatov, M. Mishin, A. Uvarov, A. Kondrateva and S. Alexandrov, "Evolution of the microstructure in titanium dioxide films

during chemical vapor deposition", *Phys. Status Solidi A*, vol. 212, no. 7 (2015), pp. 1533–1538.

[58] W. Weinreich, L. Wilde, J. Muller, J. Sundqvist, E. Erben, J. Heitmann, M. Lemberger and A. J. Bauer, "Structural properties of as deposited and annealed ZrO_2 influenced by atomic layer deposition, substrate, and doping", *J. Vac. Sci. Technol. A*, vol. 31 (2013), article number 01A119.

[59] N. V. Nguyen, A. V. Davydov, D. Chandler-Horowitz and M. M. Frank, "Sub-bandgap defect states in polycrystalline hafnium oxide and their suppression by admixture of silicon", *Appl. Phys. Lett.*, vol. 87 (2005), article number 192903.

[60] M. Seo, Y.-S. Min, S. K. Kim, T. J. Park, J. H. Kim, K. D. Na and C. S. Hwang, "Atomic layer deposition of hafnium oxide from tert-bytoxytris (ethylmethylamido) hafnium and ozone: rapid growth, high density and thermal stability", *J. Mater. Chem.*, vol. 18 (2008), pp. 4324–4331.

[61] X. Nie, F. Ma, D. Ma and K. Xu, "Thermodynamics and kinetic behaviors of thickness-dependent crystallization in high-k thin films deposited by atomic layer deposition", *J. Vac. Sci. Technol. A*, vol. 33 (2015), 01A140.

[62] W. Weinreich, R. Reiche, M. Lemberger, G. Jegert, J. Muller, L. Wilde, S. Teichert, J. Heitmann, E. Erben, L. Oberbeck, U. Schroder, A. J. Bauer and H. Ryssel, "Impact of interface variations on J-V and C-V polarity asymmetry of MIM capacitors with amorphous and crystalline $Zr_{(1-x)}Al_xO_2$ films", *Microelectronic Engineering*, vol. 86 (2009), pp. 1826–1829.

[63] J.-H. Kim, S.-G. Yoon, S.-J. Yeom, H.-K. Woo, D.-S. Kil, J.-S. Roh and H.-C. Sohn, "Electrical properties in high-k HfO_2 capacitors with an equivalent oxide thickness of 9 Å on Ru metal electrode", *Electrochemical and Solid-State Letters*, vol. 8, no. 6 (2005), pp. F17–F19.

[64] S. Meng, C. Basceri, B. W. Busch, G. Derderian and G. Sandhu, "Leakage mechanisms and dielectric properties of Al_2O_3/TiN-based metal-insulator-metal capacitors", *Appl. Phys. Lett.*, vol. 83, no. 21 (24 November 2003), pp. 4429–4431.

[65] E. William Conwell III, S. W. Muir, D. A. Keszler and J. F. Wager, "Barrier height estimation of asymmetric metal-insulator-metal tunneling diodes", *J. Appl. Phys.*, vol. 114 (2013), article number 213703.

[66] M. Sato, C. Tamura, K. Yamabe, K. Shiraishi, S. Miyazaki, K. Yamada, R. Hasunuma, T. Aoyama, Y. Nara and Y. Ohji, "Cathode electron injection breakdown model and time dependent dielectric breakdown lifetime prediction in high-k/metal gate stack p-type metal-oxide-silicon field effect transistors", *Jpn. J. Appl. Phys.*, vol. 47, no. 5 (2008), pp. 3326–3331.

[67] I. Hirano, T. Yamaguchi, Y. Nakasaki, R. Iijima, K. Sekine, M. Takayanagi, K. Eguchi and Y. Mitani, "Effects of electron current and hole current on dielectric breakdown in HfSiON gate stacks", *Jpn. J. Appl. Phys.*, vol. 51 (2012), article number 041105.

[68] D. Ielmini, A. S. Spinelli and A. L. Lacaita, "Experimental evidence for recombination-assisted leakage in thin oxides", *Appl. Phys. Lett.*, vol. 76, no. 13 (27 March 2000), pp. 1719–1721.

[69] M. Ettenberg and C. J. Nuese, "Reduced degradation in $In_xGa_{1-x}As$ electroluminescent diodes", *J. Appl. Phys.*, vol. 46, no. 5 (May 1975), pp. 2137–2142.

[70] D. V. Lang, "Recombination-enhanced reactions in semiconductors", *Annual Review of Materials Science*, vol. 12 (1982), pp. 377–400.

[71] A. Matsuzawa, "Analog IC technologies for future wireless systems", *IEICE Trans. Electron.*, vol. E89-C, no. 4 (Apr. 2006), pp. 446–454.

[72] A. Matsuzawa, "Digital-centric RF CMOS technologies", *IEICE Trans. Electron.*, vol. E91-C, no. 11 (Nov. 2008), pp. 1720–1725.

[73] R. Jacob Baker, *CMOS Mixed-signal Circuit Design, Second Edition*, IEEE, New York, USA, 2009, pp. 1–324.

Index

Printed in the United States
By Bookmasters